Two Houses of Oikos— Essays from the Environmental Age

James A. Schaefer

Two Houses of Oikos—
Essays from the Environmental Age

Copyright © 2015 James A. Schaefer

ISBN: 978-1-927685-11-2

First Edition

The cover is illustrated and designed by Maya Schaefer. Cover images © Maya Schaefer.

See "Notes and References" for attribution covering articles that have been reprinted from the *Toronto Star*, *Edmonton Journal*, and *Peterborough Examiner*,

All rights reserved. No part of this book may be reproduced or transmitted in any form or by any means, electronic or mechanical, including photocopying, recording, or by any information storage and retrieval system, without permission from the publisher, except for the inclusion of brief quotations in a review.

Printed in the USA

Publisher: Moon Willow Press
http://www.moonwillowpress.com
Coquitlam, British Columbia, Canada

INTRODUCTION ... 11

ONE WORLD IN TWO HOUSES ... 13

Smaller Bites, Better Taste ... 13
The New Environmental Unconsciousness ... 17
Taxing Nature's Services ... 23
Aesop's Tortoise ... 27
Pipe Dreams and Future Fortunes ... 31
Growing to Extinction ... 35
To Be a Better Ancestor ... 39

THE DICE OF LIFE ... 45

Gambling without Spin ... 45
Are You Scientifically Literate? ... 49
Variation, Meet Variety ... 51
Your Days Are Numbered ... 57
Uncertainty? Bet on it ... 61

NATURE'S ALPHABET ... 65

The Tale of Lord and Lady, Minus a Few Characters ... 65
Moon Shot or Pie in the Sky? ... 71
Life Is a Highway ... 75
Expect Delays, in Slow Motion ... 79
A Brave New World for Birds and Beasts ... 83

THE MOST SUCCESSFUL ENTERPRISE ... 89

Good Advice Is a Science ... 89
The Measure of Things ... 93
Evidence Means a Sporting Chance for All ... 97
Of Microbes and Men ... 101
Translation Is a Tricky Business ... 105
Looking Past the Coin Flip ... 109
A Little Feedback on Your Weight ... 113

AUTO PILOT AND POINTS NORTH ... 117

Build it and They Will Bike ... 117
Meet FLORC ... 121
Caribou Century ... 125
Bring on the White, Not the Blues ... 129

ACKNOWLEDGEMENTS ... 133

ABOUT THE AUTHOR **135**
NOTES AND REFERENCES **137**

*In memory of my father,
Keith Ronald Schaefer —
outdoorsman, businessman, scientist, humanist*

Introduction

Science, declared Sir Peter Medawar, is incomparably the most successful human enterprise, ever.[1] A peek into the mirror or a glance at your surroundings seems to confirm it. We've realized monumental scientific and technological achievements within a handful of generations. In the grand list of accomplishments by our species–art and architecture, dance and drama, music and literature–Sir Peter puts science at the very top.

Science may stand apart, but there are unexpected links, I've learned, between music and models of population growth, between mammalian physiology and a Tintoretto masterpiece.[2] The ties between science and the human experience are more than mere curiosities. Scientific success is important. Its relevance to society is crucial.

And this century will test our scientific resolve. For conservation biologists like me, one question looms large: How many species will we carry with us into the future? Here is another intriguing connection–between species and language–the most fundamental unit of biology and the most human of human traits.[3] Both species and languages evolve; many are at risk. Scientists have estimated the pace of species extinctions to be 1000-times higher than what could be considered normal[4] ; linguistic scholars are forecasting an imminent decline in language diversity. UNESCO estimates that 3,000 languages, nearly half the world's total, may vanish by the end of

Introduction

the century.[5] It's a double wallop, as the loss of a language can mean the loss of intimate knowledge about biodiversity. Indigenous forms, whether languages or living things, are especially vulnerable.

Language is integral to the study of biology, too. As every student soon discovers, it's useful to know a little Greek or Latin, because a species name reveals something about the creature–or at least how we regard it. I have my favourites. The wolverine is *Gulo gulo* ("glutton"), the grey fox is *Urocyon cinereoargenteus* ("silvery ashed dog tail"), and of course, there is *Homo sapiens* ("wise man").

For ecologists, one word occupies a special place: *Oikos*. Meaning "house", it is the Greek root of *ecology*, the relationship between organism and environment. Its linguistic origins underscore the longstanding significance of ecology to people.

Of course, ecologists do not have a monopoly on *Oikos*. It, too, is the root of *economics*, that massive branch of inquiry and interest. How we manage these two houses–indeed, the extent to which we can bring them under one roof–will be telling. Our understanding of ecology and its connections with economics will be key to our prosperity, as well as the fate of countless species.

If you own a house, my father observed, you're always working on it. Now, there are some words of wisdom–particularly in the Age of the Environment.

One World in Two Houses

Smaller Bites, Better Taste

I have a theory about dessert. You know those sweet temptations—the slice of cheesecake that ruined your New Year's resolutions, and the chocolate that wrecked my plans to be slimmer. Oh, they're good. But despite the advice from doctors and dieticians, we succumb a little too often to sweets, like a doughnut at coffee break or cookies as late-night snack. And we all know the statistics on obesity.[1]

Here's my theory: The first and last bites are best. No matter the size of the portion, the start and finish of dessert are most gratifying. This is because our senses—whether sight, hearing, touch, smell or taste—don't respond to sameness in our surroundings. They detect changes. Our palate delights in the arrival of sweet, and we savour the last morsel. Dessert is good—especially good—to the last crumb.

And this theory applies to more than pie and ice cream. Each day, we are faced with countless choices. Some decisions seem trivial—another slice? Others are profound—time to have a baby? This principle can guide us toward better decisions, big and small. And in the age of consumption, choices are more important than ever. From TV, the Internet, radio, and billboards, we are bombarded with relentless calls to accumulate more, consume more. More is better.

But try this thought experiment: Consider the number of toilets in your house. If you have ever gone camping or lived with an outdoor privy (I did for more than three years), you know the value of a flush toilet. No more frozen seat, and goodbye mosquitoes. What about installing a second toilet in your home? Yes, your daughter is a dawdler in the shower, so another facility would be a plus. A third? You could put off that weekly cleaning chore. A fourth? A fifth? Once toilets outnumber people in your household, you see the declining benefits, even drawbacks, of another john. More is less.

This is the Law of Diminishing Return. And it applies to the value of toilets, satisfaction and income, even the number of children you have. Research shows, for example, that money buys happiness–to a point. Around the world, personal well-being rises with a nation's wealth, but above $10,000 GDP per person, pleasure hits a plateau. Once a country reaches a moderate amount of income, further economic growth does not enhance happiness.[2] And when it comes to families, parents are happier than couples without children, but the effect declines with each child. On average, the first child increases happiness a lot, the second a little, the third not at all.[3] There is science behind the proverb: Happiness is the comfortable space between too little and too much.

Even money, despite our insatiable appetite for it. Consider shopping for a number of items you could purchase for $20 or less. If you ranked these things from most desirable to least, $200 would buy you the top 10 things on your list. With another $200, you would start buying less desirable items. The more money you have, the less each dollar is worth to you. And this is the basis for a progressive tax system. A $2,000 credit is likely to mean more to the worker on minimum wage than $50,000 to the CEO with a seven-figure salary. The poor represent a better return on investment.[4]

The two-bite brownie is more than a light snack. It is an idea—more fulfilment for each of us and a fairer society for all—if we can resist devouring the whole bagful. The best desserts come in small packages.

THE NEW ENVIRONMENTAL UNCONSCIOUSNESS

It was the sigh of relief heard 'round the world–the beginning of the end of the Fukushima nuclear crisis in Japan. One month after the calamity began, the radiation leak from the crippled reactors was sealed. But it will take longer to heal the economic and ecological wounds. The stark images will linger, too. For this generation, the word Fukushima will be shorthand for environmental tragedy: earthquake, tsunami, contamination, evacuation.

To many people, the near-meltdown was another grim tale in our troubled relationship with nature–to be placed alongside Chernobyl, the Exxon Valdez, and the Deepwater oil spill in the Gulf of Mexico. But environmentalism is changing. Quietly, in research labs around the world, environmental thought is taking a new path. And these discoveries are happening very close to home: Not in the abyss of the Gulf, along the distant shores of the Pacific, or in some other faraway place–but in the depths of the human mind.

This is the green effect. It's how nature colours our thinking, often in subtle and surprising ways.

You can find part of this new environmental frontier in the American Midwest. A few years ago, families in Urbana-Champaign realized something was curiously right. Parents in this Illinois city were reporting that their children, diagnosed with ADHD, were unexpectedly better behaved. Their kids seemed soothed and more focussed when their after-school activities took place in greenspace, like a neighbourhood park.[1]

ADHD–attention deficit hyperactivity disorder–is the most common mental health problem among children in North America, afflicting about one child in twelve. Plagued by inattention and impulse, kids with ADHD are at risk of depression, failure at school, and rejection by their peers. Families feel the strain. Could managing ADHD, literally, be a stroll in the park?

Science turns a notion into an experiment. Psychologists at University of Illinois, Andrea Faber Taylor and Frances Kuo, designed a test.[2] They arranged for pre-teens identified with ADHD to be taken on guided walks: through a city park, quiet sections of downtown, or a residential neighbourhood. Afterwards, a researcher, "blind" to which setting the child had experienced, administered Digit Span Backwards test–a measure of concentration. A subject listens to a string of digits, then repeats the sequence aloud in reverse order. Question: "4-1-6?" Answer: "6-1-4." With each correct response, the sequence is lengthened, to as long as eight digits.

The result was striking.[2] Following the stroll in the park, kids concentrated better than after walks in less-natural settings–roughly equivalent to the peak effect of a dose of extended-release methylphenidate. (You know this drug as Ritalin.) Here is a promising treatment: safe, inexpensive, and free of social stigma. *Whatever.* To the kids, the park was simply more relaxing and more fun.

The study backed up everyday experience. One parent told the story of his son, regularly struggling with his attention deficit symptoms: Yet he "hit golf balls with me for 2 hours at a time" and he "fishes for hours" after which "he's very relaxed."[1]

How does nature seep into our minds? In the late 19th century, psychologist William James proposed that elements in the natural environment are easily engaging: "strange things, moving things, wild animals, bright things, pretty things"—a tangle of intrigue. Relaxed and absorbed by nature, our brain is restored and ready to take on new, demanding tasks.[1,3]

And according to theory, the more intricate the environment, the greater the effect. In Sheffield, England, another research team revealed the experiences of more than 300 park-goers enriched by nature's complexity. Their ability to think increased with more varieties of plants, birds, and habitats.[3] (The harried worlds of shopping and TV, in contrast, seem to provide no such restorative effects.) This is the irony of greenery: Clarity of mind from the clutter of nature.

In perhaps the most striking example, Dr. Kuo and her colleagues linked vegetation to urban crime.[4] Their focus—a housing project in inner-city Chicago. Outside otherwise identical buildings, there were big differences in greenery. Some had shaded, grassy courtyards; others had barren expanses of dirt or concrete. Residents were similar in income, education, and employment, and they had been randomly assigned to their apartments. It was an ideal, controlled study.

And the impact was surprising. Residents living next to trees and grass reported lower levels of fear, and less aggressive and violent behaviour. Police reports showed fewer property crimes and fewer violent crimes. These results debunk the myth that green spaces promote crime by providing concealment for lawbreakers. Indeed, the opposite– grass and canopy trees invite residents out-of-doors, where they get to know their neighbours and where they provide more "eyes on the street".[4] Trees can be tough on crime.

The green effect has been uncovered in other unexpected places: In hospitals, where patients healed faster and required less medication after surgery when given a view of trees instead of a brick wall.[6] On campus, where college students got a boost in cognitive performance after walking in nature compared to downtown.[5] Along roadways, where drivers stuck in traffic, but surrounded by attractive nature, experienced lower stress than those mired in dense, built-up areas.[6] Who would have suspected greenspace as an antidote to road rage?

A final twist: Just caring about nature is linked to quality of life. Research shows that having an attachment to positive environmental features, like landscapes and wildlife, is associated with one's well-being.[7]

A crisis like Fukushima reveals our usual, one-way view of the environment–concern about our multitudes, our wastes, and our global resource demands. But there is a subtle flip side. It's about how nature affects us.[8] Environmental unconsciousness, it seems, is deeply ingrained—in ways more positive and profound than we knew.

TAXING NATURE'S SERVICES

T4 slips and RRSPs—for Canadians, these are the distant early warnings of tax time. As my wife and kids will tell you, the complexity of tax returns does not bring joy to our household.

But imagine this: A $7,000 windfall for you and every member of your family—not just this year, but each and every year. An enticing proposition, you say, although surely this is just another vague election promise, bound to be broken.

This windfall, however, is delivered and underwritten by Nature. The payoffs are "ecosystem services"—the goods, services, and enjoyment that we derive from the natural world, everyday, for free. As beneficiary, you receive 17 no-charge services including purification of water, regulation of greenhouse gases, and mitigation of droughts and floods.[1] Only one proviso—that nature is admitted the space and species to do her timeless task, to supply these benefits to you.

Your take is part of $47 trillion—the estimated annual value of the planet's natural capital, a colossal figure that surpasses the worth of the whole human economy.[2] Even so, critics have called it a serious underestimate ... of infinity. Should we foolhardily destroy an irreplaceable life-support system, they caution, no one will be left to collect the 47 trillion, whatever the currency. But putting a price on the benefits of nature is instructive. It conveys immense worth, otherwise neglected in our dollar-driven world.

Still, such a whopping sum is not easy to grasp. So let's focus on one piece of the story: the birds and the bees. This is not some wild tale about reproduction. It is about pollination–and therefore it includes you. Pollination is essential for flowering plants to produce seeds and fruit and, as farmers know, it is crucial to food production, for crops like potatoes, almonds, soybeans, oranges, peppers, and canola. In North America, it is estimated that one mouthful in three is food that can be traced to animal pollinators like birds and bees, but also bats and moths. One in three–now there's a morsel worth digesting.

If pollination is vital, how might we enhance it? A study of watermelon production in California reveals a simple answer: Habitat is key.[3] Native bee species can provide all the pollination needs on the farm if upland habitat is close by–if 40% of the vicinity is retained as oak woodlands or chaparral. Even just 10% of the land as habitat will provide about one-third of a farm's pollination requirements. Similar results have been shown for canola and sunflowers–improved yields if we set aside some wildlife habitat.

The news gets even better. Not just the amount of pollination is improved by upland habitat, but also more consistent rates of pollination[3]–a sort of insurance policy, in the form of crop yield stability, for the producer and consumer. Wildlife habitat means resilience for the farm and security for the food supply. What's not to like?

But we are removing habitat at an unprecedented rate and this is the primary threat to species, nationally and globally. And as we put ecosystems under strain, we risk undermining of our own prosperity.[1] Perhaps this is the price of progress. In our just-in-time society, there is little place for farsightedness, for reflection on what is best in the long run. So we press forward, dutifully accepting that species loss is unavoidable. "We continue to go boldly where no one wants to go", in the words of economist David Korten.

The benefits of habitat conservation invite us to chart a new course. The gains are bigger than we imagined–not just for nature, but for ourselves. To safeguard wildlife is to safeguard the future.

Aesop's Tortoise

In the year to come, I boldly predict that you'll accomplish the improbable. You're going to complete a marathon—considered the most gruelling of endurance races—without training, doping or hardly even trying. Whatever your current form, I assure you this feat can be accomplished before the end of the calendar year. No sweat.

A marathon is the pinnacle of distance events— 42.195 kilometres. Simple arithmetic shows that to finish a marathon by the end of the coming year you'll need to walk a mere 116 metres per day starting January 1st—a daily jaunt about the length of a football field. Heck, you could complete this distance with a routine stroll to the corner store. If a somnambulist, you might do it in your sleep.

But you object, "That doesn't qualify!" Sure, a marathon in 365 days doesn't approach the Olympic standard, but what distinguishes the dedicated distance runner from the rest of us can be summed up in one word—rate. Virtually everybody can complete a marathon in a few months; only conditioned athletes can do it in a few hours. And no human has yet performed the feat in less than 2:02:57, the world record.

Pace is important—it's what earns a trip to the podium—but it's not restricted to physiological capacity. Rates are crucial to other aspects of biology, like ecology and evolution. And problems can arise when rates clash, especially mismatches between natural systems and people, like the tempo of our environmental impacts or societal expectations.

Take climate change. Birds and butterflies tell us that a warmer world is already here. Bit by bit, myriad species have been making poleward shifts in their distribution—an unmistakable signal of a hotter planet.[1]

And this is not the first time the world's biological map has been redrawn. At the end of the last ice age, glaciers receded and plant species extended their ranges in step with the retreating ice. Typical northward advances by trees, like pine and spruce, were 100 to 400 metres per year—a floral procession at a glacial pace.

It's not just the magnitude of global warming that might prove unpleasant, but its swiftness. A 3-degree Celsius rise this century would amount to change 33 times faster than since the last glaciation—equal to the disparity in running speed between a human and a crab. Some species may not be able to keep pace.

This upheaval is likely to fracture communities of living things. It's as if your whole town had to relocate in a string of uncoordinated departures, with no assurance of safe arrival. Imagine teachers, police, garbage collectors, movers and construction crews arriving at the new location in disjointed fashion, or perhaps not enduring the journey at all. Intimate connections would be severed; services would be lost.

Of course, no species lives forever. The fossil record reveals that most organisms that have ever lived have gone extinct. It's the tempo of impoverishment that makes our times unprecedented.

Biologists estimate at least 3,000 species are vanishing each year, easily outstripping the appearance of new life forms at about one new species per year.[2-4] Even in our day-to-day lives, the effects of speed can be counterintuitive, sometimes counterproductive. Drive faster and arrive sooner, right? A traffic study from the U.K. suggests not. If the speed along a multi-lane expressway is doubled from 50 km/h to 100 km/h, it can handle only one-third as many automobiles. At 110 km/h, the capacity declines further, to just 27 per cent.[5,6] The benefits of speed are overwhelmed by the need for safe stopping distances. Slower is faster.

And slower might be the key to improving our environmental prospects. There is encouraging news that nature is resilient, given the chance. Climate scientists tell us that limiting global warming to 0.1 degrees C per decade would diminish the risk of unwelcome surprises. And living resources–from the salmon on your table to the wood in the table itself– are renewable, capable of replenishment. Wealth in the long haul means tempering our demands to match the speed of nature.

Like Aesop's fable of the tortoise and the hare, arriving at a prosperous future is not a foot race. It requires a measured pace. At the end of the day, a sustainable outcome would represent the most prized gold medal of all.

Pipe Dreams and Future Fortunes

Each summer is different. This year, we have been dogged by drought, but also by controversies that won't melt away: Pipelines, pipelines, and pipelines. These mega-projects are touted as vital for delivering Alberta crude to market. If oil is the lifeblood of an energy superpower, pipelines are the arteries. It is the new National Dream.

And a bonanza in the zillions. Together, the Keystone XL, Northern Gateway, and Trans Mountain projects will cost $15 billion, promise billions more in revenue and, every day, transport 2 million barrels of oil sands bitumen[1]–equal to the daily volume flushed down the toilet by 11 million people. The numbers are big, but the math looks elementary. Oil out, jobs and money in. Perhaps this is why the Prime Minister called approval of the Keystone XL project a "no-brainer."

But there is another side to this ledger–water. Some of the continent's most magnificent and essential waterways, like Queen Charlotte Sound and Ogallala Aquifer, are set against the oily reminders of environmental mishap, like the Exxon Valdez and Kalamazoo River. Pipeline proponents and opponents appear worlds apart. Indeed, the two sides seem to mix as well as oil and water.

The paradox is that these opposing views are not really about oil. They are about time.

And about choices. As an illustration, consider a mathematical choice of fortune. Imagine for a second that you, winner of a lottery grand prize, must select one of two payment options: A lump sum of $1 million, or a series of payments—one penny the first day, double that penny the next day, then double the previous day's pennies and so on for a month. Which option would you choose?

The answer is no small change. At end of the month, 'double the penny' gives a return of $10,737,418.23—more than 10 times the winnings. Even February, shortest month of the year, leaves you $342,000 richer than the lump sum.

Most of us, knowing the answer, would be patient to reap the bigger reward. And this is the moral of the numerical tale. It's in our interest to think long term. Yet we seem hard-wired to seek instant gratification, even when counter to our well-being. Think of smokers wanting to quit, workers wanting to save for retirement, cravers of fast food wanting to lose weight. Resisting impulse might be one of the biggest challenges in our lives.

And so it seems with pipelines. Despite the rush, many economists see a steady climb in oil price—driven by shrinking supplies and insatiable demand, especially from developing countries. Indeed, almost every year since 2001, the price of crude has risen, quadrupling in a decade, easily outpacing inflation. Expensive oil looks like a permanent fixture for the future. This implies that oil-in-the-ground in a few decades will be worth much more than oil-in-the-pipeline now. And perhaps, upon careful reflection, we may decide safeguarding water means even more. Wisdom is taking time to consider the options.

It is an enduring tale. In *Storybook Dictionary*,

children's author Richard Scary poked fun at the perils of nearsightedness. We learn 'H' is for hole. Haggis the cartoon dog hasn't repaired the thatch of his little house:

"Haggis has a hole in his roof.
He never fixed it because on rainy days it is too wet to work.
And on sunny days it doesn't need fixing."

Our economy cannot prosper in a failing environment–our home, in its broadest sense. We simply need to have foresight, to act wisely, especially when the sun is shining.

GROWING TO EXTINCTION

When a New Year is at hand, so, too, is the perennial ritual of forecast and resolution. What lies ahead? One item dominates annual projections: Growth. Savants and soothsayers will predict growth of the economy, greater personal fortune and, despite our best intentions, expansion of waistlines. But beyond the finances and food, we notice some wear and tear. Wild species are vanishing.[1] Our economy and ecology seem to be heading in different directions.

This apparent rift between economics and ecology strikes close to home—literally. Both words are derived from the Greek, *oikos*, meaning "house". The common etymology underscores the importance of both economy and ecology for human well-being.

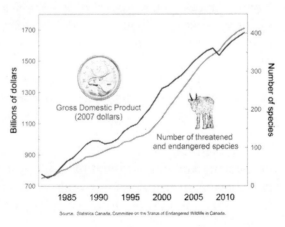

Figure 1. Growth in GDP in Canada (in 2007 dollars) and in the number of threatened and endangered species, 1981-2013.

But the relationship is not just a matter of speech. It is cause-and-effect. Since the early 1980s, the Canadian economy has expanded almost non-stop, predicated on increased production and consumption of goods and services. We have benefited handsomely, but as we consumed more, the number of threatened and endangered species has also grown[2]–almost 10% per year (Figure 1, previous page). The tally of species-at-risk has been doubling every 7 years. This is a downside of up.

Perhaps these numbers are inflated. Economists can adjust values to a constant currency, like 2007 dollars, but biology is also prone to inflation. Using new tools like genetics, we occasionally discern two species or populations where once we saw just one. This biological inflation, however, cannot fully explain the dramatic increase in imperiled wildlife. In North America, for instance, 17 species of large mammals–like caribou, wolverines, and grizzly bears–have lost more than 20% of their traditional range.[3]

Why are species declining? Simply stated, ever more resources for *Homo sapiens* means less space for wildlife. According to a Concordia University report, loss of habitat is the overwhelming reason–a driver of imperilment for six out of every seven species-at-risk in Canada.[4] Habitat degradation is more serious than pollution, alien species, and overharvesting. Globally, too, habitat loss is the principal threat to birds and mammals.

The relationship between ecology and economy

is sometimes tense, but it is not equal. The two systems differ in pace. Whereas the economy seems to gain speed each and every year, ecosystems operate at limited rates. Populations of pine trees or salmon, for instance, have an upper biological capacity to respond to harvesting. Theory indicates that slow processes constrain fast ones, implying that ecology is a constraint on our economic desires.[5,6] Or as Gaylord Nelson, former US Senator, put it: "The economy is a wholly-owned subsidiary of the environment."

The ecological-economic link may be tightening. A few years ago, for example, the Stern report predicted severe consequences if we choose to ignore global warming. Sir Nicholas, former chief economist at the World Bank, estimated that stabilizing greenhouse gases will cost about 1% of global GDP each year. On the other hand, should we choose to do nothing, by 2050 runaway climate change would cost between 5% and 20% of annual GDP.[7] And in Australia in 2007, the worst drought in a century has a wielded a 0.7% cut to economic growth.

Must ecology and economy be a house divided? Won't paying heed to the environment "hurt the economy"? Here is a false choice.[8] The true alternatives are short-term exploitation and long-term prosperity. Sustainability means a focus on enduring wealth, especially for future generations, rather than mere personal and immediate gratification. At its most fundamental level, it means broad thinking. As historian Donald Worster noted, "A wider area of responsibility ... is all that conservation asks."

Both ecology and economy are central to our well-being. This year, to benefit ourselves as well as future citizens, we might resolve to put our whole house in order.

TO BE A BETTER ANCESTOR

Running is an odd activity. To non-enthusiasts, it is a mix of monotony and self-inflicted pain. Yet, in Canada, many adults list running as their favourite participation sport–more than soccer, softball, hunting, fishing, aerobics, or tennis.[1] Strange pleasures, it seems, are found on the trails and treadmills of this country.

I run–sometimes. To be honest, my pace is much closer to a plod. On top of the soreness and stiffness, I can recount incidents of slipping on ice, blisters, dog bites, overheating, and freezing. *Oh joy*. This is the gift of experience: a long list of excuses for not lacing up the shoes. Not today.

I've also discovered something more detestable than running: not running. Indeed, many of us have mixed feelings about exercising and other personal resolutions. Time is the nub of the problem. Workout or diet, anyone? The discomfort is assured, obvious, and immediate. The upside, if any, appears somewhere down a long and uncertain road. The pains, the gains are doled out at different times– weeks, months, even years apart.

For many, this calculation is way too easy. Impulse wins out. But when the time horizon reaches decades or centuries, the computations become prickly. What is the value of something distant? The answer is pivotal to the future–and not just your future. Especially not yours.

A choice experiment is one way to get a handle on people's preferences and how they change with time. Start with a straightforward question.

Question 1. Which do you choose: $100 now OR $100 one year from now?

That's a snap. Almost everybody would pocket the instant cash for some very simple reasons. We are mortal beings and there's some doubt about that future payment. Even though $100 hasn't changed, what has changed is our perception of $100, coloured by time. A benefit delayed is a benefit diminished.

Let's change up the numbers. Another question.

Question 2. Which do you choose: $100 today OR $200 one year from today?

Smaller sooner or larger later? Studies show that most respondents still go for the fast cash—an inclination to snatch the immediate reward, even at the expense of a bigger, future reward.[2] This behaviour also reveals the rate at which people devalue a future reward. Economists call it the discount rate. A high discount rate seriously diminishes the future value; a low discount rate places the future value on par with a current one.[3]

Welcome to Discount World. More than just dollars are at stake here. Many species—the passenger pigeon is the most famous example—have been harvested to extinction. Today, around the world, over-exploitation remains the third most common reason for the decline of wildlife. Consider the blue whale, the largest animal ever to roam the globe. An adult blue can stretch to one-third of a football field, weigh the equivalent of 13 city buses, accelerate to 50 kilometres per hour, and live for more than a century. Here is a fascinating creature—bigger than a yacht, faster than a yacht, more durable than a yacht.

And, in the past, more valuable than a yacht. During the 20th century, 369,000 blue whales were taken in the Southern Ocean—the most stunning case of wildlife exploitation in history.[4] Propelled by steam engines, explosive harpoons and short-sightedness, harvesters pushed these whales to the brink, to less than 1% of their original numbers. Whaling is now banned. Ethics aside, this brush with extinction seems bizarre. Surely a more measured pace—a sustainable harvest—would have been more profitable?

Think of blue whales as offshore banknotes, cashable at any time.[5] Choose one of the following options: (A) Harvest these whales sustainably and indefinitely or (B) Promptly liquidate the stock and re-invest elsewhere. While a sustainable harvest looks sensible (roughly 5% annual return), a larger payoff comes from aggressively harvesting to extinction and re-investing in a higher yield, dot-com or other stock. This is a stark and sobering conclusion. Extinction is entirely rational from a dispassionate, economic standpoint.

Not surprisingly, some creatures cannot thrive in Discount World. Large-bodied animals, like blue whales, have a "slow" lifestyle. They are late to mature and slow to reproduce; an average mother is more than 30 years old and she gives birth to a single calf only every 2-3 years. The financial return—how quickly the species can reproduce in the face of harvesting—is no better than 7% per year. These biological traits are shared by countless plants and animals: sharks, tigers, parrots, rhinoceros, elephants, chimpanzees, sturgeon, white pine and many other timber trees. Little wonder these species are vulnerable to over-exploitation. They are simply too slow to be included in our portfolio.

In Discount World, the economic value of wildlife in the long run can fade to zero. It is a faulty telescope. The tale of the blue whale is also a reminder that discounting cannot be divorced from ethics.[6] It is not merely a question of dollars, but sense.

Given the choice between $100 now and $200 in one year from now, many people preferred the instant reward. Let's introduce a subtle twist—with a surprising result.

Question 3. Which do you choose: $100 in five years OR $200 in six years?

You see the change. The situation has shifted to the future. And with it, most people accept the delay and favour the larger, $200 payoff.[2] This answer is telling. When we visualize our better selves, our future selves, we stress the long term. We are more patient. We discount less. Time changes our perceptions of time.

Humans are intriguing creatures. Among animals, our species is unrivalled in its capacity for language, for manufacturing and using tools—and for anticipating long-term future events.[7] With little effort, we can picture succeeding generations. Answers become vivid. Ponder some of the masterpieces of nature—the Great Barrier Reef, Great Lakes, and Great Blue Heron. Won't these natural treasures be as important to your great-grandchildren as they are to you? [8] The virologist

Jonas Salk posed the key question: Are we being good ancestors?

To be good and fair ancestors, we will need a variety of tools, including financial levers, applied with care and foresight. These days, a growing chorus of economists and scientists is calling for a price on carbon–a simple economic instrument to reduce greenhouse gas emissions, encourage alternate energy technologies, and lessen the hazards of climate change.[9,10] A new tax, even if revenue neutral, may not be warmly welcomed. But we might listen to the future–and perhaps realize there's something far worse than a carbon tax: No carbon tax.

Decisions, from planetary to personal, are a feature of modern life. I'm working on a new philosophy to continue lacing up the shoes and keep running. It's one stride at a time, eyes firmly fixed on the horizon.

The Dice of Life

Gambling without Spin

Sport, they say, is a window on the soul of a nation. And few events can match the spectacle of a Stanley Cup final–the crucible-on-ice, fueled by spectators and huge TV audience. But the build-up takes months. Rounds of punishing playoffs follow an exhausting season and pre-season. But why not skip the preliminaries and go straight to the drama of sudden-death? One game. First goal wins.

Few would favour such an abrupt conclusion. Apart from the loss of revenues and suspense, much would be left to caprice–a lucky bounce, a questionable call, a fluke goal. Lord Stanley demanded better. The Cup is not awarded in a simple shoot-out, nor is the World Series decided after one inning, Wimbledon after one set, a golf championship after one round. We prefer the long contest–a sporting chance to allow skill to prevail. "The longer you play", said Jack Nicklaus, "the better chance the better player has of winning."

If sport is a window on our collective soul, gambling could be a glimpse of our reasoning–the two distinguished by skill versus chance. And the king of the casino is the slot machine, amassing more income than all other forms of gambling combined. Today, the mechanical one-arm bandit has gone digital, capable of flashy graphics and big takes. A slot machine can vacuum up $500 a day–an impressive six-figure annual income.

Such enormous gains seem surprising, given that slots in my home province of Ontario, for example, have a regulated payout of 85% or higher. In other words, with a 90% payback, a player can expect to enter with $100, leave with $90. The edge for the house looks meagre, but it's deceptive. A typical gambler may buy $50 or $100 in tokens and continue playing, as the machine gnaws away, until the money is gone. Any winnings end up back in the machine. On average, $100 will be whittled down to $35 after just 10 rounds. To be cleaned out, a player need not be unlucky, just persistent. In the long run, the house always wins.[1]

But some players are convinced that outcomes, at least in part, are due to skill. For example, some cling to the superstition, after a string of losses, the slot machine must be "due". This is the gambler's fallacy–that a run of luck, whether good or bad, influences future chances.[2,3] But the machine has no memory. Each spin is independent. The upshot is that some gamblers, rather than cutting their losses, sink deeper into debt, banking on a big but elusive win, an illusion to erase debts and cure problems.

And, although superficially indistinguishable, slot machines differ in payback percentage–from 85% to 98%. Studies confirm that many players can discriminate between "loose" and "tight" machines. But this skill, too, is illusory and carries the added danger, especially for problem gamblers, of overestimating their abilities. No slot machine generates consistent wins.

Gambling is dicey–and not just for players. The

estimated number of problem gamblers in Ontario is immense, equivalent to four times the population of the city of Peterborough. And it carries risks of absenteeism from work, family neglect, divorce, even violence against partners. The societal price, too, is massive. In the US, annual health and employment costs from gambling troubles are some $5 billion, with another $40 billion in lifetime costs owing to bankruptcy and legal burdens.[4] These are not happy numbers.

Our national self-image is intimately tied to sport, along with the skill, determination, and fair play it demands. Gambling is a matter of chance—but its presence in your community is not blind luck. It is for you to decide.

ARE YOU SCIENTIFICALLY LITERATE?

Now you can find out. Physicist and author James Trefil of George Mason University devised a short quiz, published in the *Toronto Star*.[1] There are 26 multiple-choice questions on biology, physics, and chemistry. Score 80% and you make the grade, according to Professor Trefil.

(I admit to apprehension about taking the test. But I can state, happily, my score met the standard for literacy–a reassuring result for a career scientist.)

Of course, multiple choice exams have their drawbacks. With four choices per question and no penalty for an incorrect answer, a know-nothing could expect a mark of 25%–still mired in 'F', but better than a goose egg.

For the indecisive and the unlearned, however, there are more strategic methods than sheer guessing in multiple choice tests.[2] One ploy, when in doubt, is to choose answer (c). Examiners seem to find this letter a favourite, presumably to conceal the correct answer amongst the wrong ones. Another trick is to pick the longest answer, given that teachers tend to add details to make the correct answer entirely true.

Adhering to the "choose (c)" rule, I scored 42% on the quiz–significantly better than random.[3] Choosing the longest answer gave an even more impressive grade of 58%–clearly superior to random picks[4], but still short of true literacy. Alas, it appears there is no substitute to learning science to become scientifically literate.

VARIATION, MEET VARIETY

Economy. Volatility. More than ever, here are two words uttered in the same sentence, commanding the attention of Presidents and Prime Ministers. A roller coaster economy is the ride of the times. And with one ticket, it delivers the best and the worst–fleeting wealth and opportunity, alongside lost savings and lost jobs. But ups-and-downs also reveal truth about ourselves. We crave stability. We design our surroundings, even change our ways, to avoid the unexpected.

Taming variation is not confined to the halls of political and financial power. A steadying hand is so pervasive that we overlook it. On the next visit to your neighbourhood bank, for instance, take note of the customer line-up: Many tellers, one queue, each person proceeding in order to the next available server. This single queue system is so widespread–at airports, ticket counters, government service offices, Tim Hortons–that surely it must cut wait times. After all, few things are as miserable or infuriating as a line-up.

The surprise is that a single queue doesn't move any faster than multiple queues. Instead, the system is designed to smooth out the fastest and slowest transactions and servers. It reduces variation–and the chance you will squander your lunch hour in line. Indeed, even more than a long wait, we detest variation in waiting–5 minutes this time, 55 minutes the next. Studies show that uncertain waits are more aggravating than waits of known length.

Variation, Meet Variety - 52

You drive home. Of course, you've buckled your seatbelt, checked the rear-view mirror, insured your car. Speed kills and so you stick to the limits–well, most of the time. A speeding ticket serves as a stinging reminder for those who don't.

But going fast isn't the whole story behind car accidents. Research shows that the risk of a crash rises when drivers stray from the average speed, whether faster or slower.[1,2] On expressways, for example, slow vehicles are rare: Roughly one vehicle in 700 travels at less than 60 km/h. But dawdlers are especially prone to accidents, accounting for nearly 10% of collisions. The culprit is variation–a hazardous mix of driving speeds. And so the Highway Traffic Act prohibits unnecessarily slow driving. Some experts advise: When in traffic, go with the flow.

Home at last. Nearing the doorstep, you easily climb the stairs. Without looking, your feet clear each step by as little as 35 mm.[3] The critical feature is their uniformity. Indeed, building codes stipulate each stair differ by no more than 3/16" in height. And with good reason. Falls on stairways injure tens of thousands of people and cost billions of dollars each year. In similarity, there is security.

This is the modern world, where the best surprise is no surprise. We order our surroundings for sameness and stability–from expanses of suburbia, to climate-controlled offices, to calls for majority government. Randomness equals anxiety. And unforeseen, even imagined perils can drive us to extraordinary measures, such as tough-on-crime laws and wars on terror. Perhaps it's a hardwired human trait. Animals like caribou appear to select habitats for their consistency. Lactation, the very feature that defines mammals, may have evolved to help mothers deal with an unreliable food supply.[4] Of course, there is a still place for variation. We seek a little excitement—within limits—like extreme sports and the big lottery ticket. A good writer know the importance of changing sentence style and structure. And a speech delivered in monotone is, well, monotonous.

But there is a paradox. In the quest for security through sameness, we can invite surprise. Residents of my city of Peterborough, Ontario, will not soon forget the great flood of July 2004. The memories are stark—in just a few hours, a whole summer's worth of rain and the sorry aftermath of sewage, closed businesses, and a city in a state of emergency. The most severe damage was downtown, where surfaces are nearly impervious. Indeed, cities homogenize the environment with asphalt and concrete, leaving little greenspace and few species. By literally paving them away, we exacerbate extremes. The asphalt city is a brittle landscape.

Of course, it is not possible to link any one event to global warming. But scientists consider extreme events to be the sharp stick of climate change. And worldwide, the evidence is mounting—more extremes in precipitation, wetter winters and drier summers, and a rise in record-breaking temperatures.[5,6] Although a few degrees of warming may sound appealing to winter-weary Canadians, we are unlikely to stroll into a new, warmer world. It's climate disruption, and pumping carbon into the atmosphere moves us closer to unwanted tipping points. As the late climatologist Stephen Schneider cautioned, we are loading the climate dice—increasing the odds of dangerous climate change. And extremes may be a glimpse of what's in store. In Peterborough, the 2004 flood took place just two years after another drenching. One downtown sign read: "I'm getting used to these once-in-a-century storms."

How can we brace for future shock? To deal with variation, paradoxically, we need varieties. Being resilient means a big toolbox—whether diverse landscapes, rich collections of species, or a plurality of perspectives. It's a common wisdom in design. From healthy eating to financial assets to metal alloys, strength and resilience are rooted in diversity. Indeed, humans may always have depended on variety. In prehistoric societies in Europe, for instance, a diversity of big game ushered in prosperity and population growth.[7]

Diversity lends security. Battling climate change, therefore, means not just reducing greenhouse gas emissions, but creating an agricultural mosaic. Research shows that setting aside a little land for conservation captures more carbon than producing corn-based ethanol.[8] In lakes, large assortments of plants and animals help to withstand acid stress as species serve in backup roles to others.[9] Variety can even represent a model for innovation and productivity when leaders in diverse roles work cooperatively. It's the art of thinking independently—together. As author Walter Lippmann observed, when we all think alike, no one thinks very much.

Life is an unending current of change—the one constant we can rely on. It was Greek philosopher Heraclitus who said that one cannot step twice into the same river. And in this steady stream of change, we can immerse ourselves, even rejoice in the flow, while keeping an eye out for any coming swell. Diversity can help us stay afloat. It is a lifeline to health and security—in short, some very 21st century issues.

YOUR DAYS ARE NUMBERED

Not long ago, reading and writing were the hallmarks of literacy. But society is shifting in a digital direction. Government surveillance of US citizens' phone records is just the latest revelation of how masses of information can be exploited. Future prosperity, even democracy, could depend on citizens having a solid understanding of numeracy. Move over, math phobia.

At this new frontier is data mining—sifting for nuggets of knowledge in heaps of information. Credit card companies, for instance, are interested in distinguishing risky customers from reliable. A classic tale concerns J. P. Martin, executive at Canadian Tire. After analyzing every transaction using the store credit card, he found certain purchases were good predictors of credit risk. According to a *New York Times* article, cardholders who bought bargain motor oil or a chrome-skull car accessory were more likely to skip payments than buyers of birdseed or felt pads for furniture.[1] Numbers tell a story, if we learn the language.

The idea is not new. In the 18th century, Joseph Butler proclaimed probability as the guide of life.

It is time to reveal your own numerical literacy. Try this test.

Question 1. Two of the statements below are dubious, based on self-evaluations. Which one is true?

(A) 88% of drivers are better than average

(B) 94% of professors are better than average

(C) 99.8% of people have more thumbs than

average.

Exaggeration and self-evaluation go hand-in-hand. Nearly one in six teenagers, for instance, believes he or she will become famous[2]–which would give rise to millions of Justin Biebers. Overconfidence affects motorists, too, who overstate their driving skills and professors who overrate their teaching abilities.[3,4] (I confess to a double-fault here.) The old maxim remains true: Know thyself.

And know that averages can be deceiving. A fraction of the population has lost upper limbs or digits from accident or disease, so the mean number of thumbs per person is around 1.998 . Most of us are above average, but that's not deserving of a thumbs-up. Next question, please.

Question 2. In any group, there's a chance at least two persons share the same birthday (month and day). How many people are needed for a 50% or better likelihood of a match?

(A) 23 people
(B) 88 people
(C) 183 people.

Coincidence is not as improbable as you think. A birthday match is likely in a small group–just 23 people. With 50 people, the probability exceeds 97%. This is because as the room fills, the number of pairs quickly rises.[5] But beware: The answer depends on the planet. On Mars, where a longer year (669 days) means more birthdays, 31 Martians are needed for 50:50 odds of a match. (I leave you the larger question of whether such an alien gathering is likely.)

Question 3. Seated at the roulette table, you

notice five consecutive spins of red. On the next spin, the chance that black turns up is:
 (A) more likely than red
 (B) less likely than red
 (C) equally likely as red.

Streaky behaviour, while catching to the eye, may result from mere chance. In 100 roulette spins, for instance, there is a good chance (73%) of five or more consecutive reds. And
the mechanics of a fair roulette wheel means equal likelihood of black and red, no matter the previous outcome. The little ball is oblivious to history.

That's our test. Congratulations if you scored 3/3. Of course, you might have answered randomly, by rolling dice or flipping coins. Straight luck would offer a reasonable chance (1/27) of obtaining a perfect grade. But you already knew that.

UNCERTAINTY? BET ON IT

The bet is on. In a public wager with two Russian climate skeptics, British climatologist James Annan staked $10,000 that global temperatures will rise over a 20-year period.[1] We'll know in just a few years—a tidy sum in the bank and scientific reputation intact.

This is the long bet, spanning years, even decades. It's befitting our times. The word "uncertain" has appeared more frequently in the news during anxiety about the economy, politics, and environment (Figure 2).

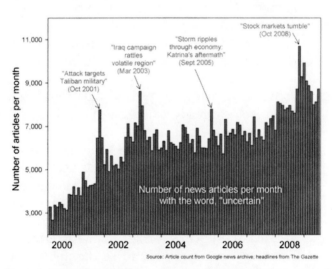

Figure 2. A monthly tally of news articles with the word, "uncertainty", during the first decade of the 21st century.

But we welcome the unknown. Every day, we make routine decisions about ourselves and our families without the benefit of ironclad knowledge. We carry umbrellas and insurance to guard against future outcomes, large and small. We may seek advice from experts, like meteorologists and physicians. Sometimes we rely on intuition.

Hunches, however, can fail us. For instance, an astounding proportion–one in ten Canadians–is banking on lottery winnings as a retirement nest egg. Buy a 6/49 ticket every week, statisticians tell us, and on average you can expect the jackpot roughly once every 250,000 years.[2] A hotter climate provides a better bet. Dr. Annan is confident of collecting the prize. As he told the journal, *Nature*, "A pay-off at retirement age would be a nice top-up to my pension."[1]

The climate wager also unmasks a deeper truth– that science is steeped in uncertainty. All scientific ideas, even widely accepted theories of evolution and continental drift, are provisional. This means that scientists are ready to revise these ideas in light of new evidence.

Take the extinction of dinosaurs. How did it happen? Most scientist trace their demise 65 million years ago to the impact of a massive asteroid. But a recent report suggests the theory needs refinement.[3] Palaeontologists discovered the Chicxulub crater in Mexico, presumably formed by the asteroid collision, is 300,000 years too old to be the impact site. New evidence leads to new research–the possibility of two or more impacts.

Science, therefore, is tentative. But it would be a mistake to think scientists revisit basic theories every morning. Some ideas are firmly established; others are more speculative. The goal of science is to reduce uncertainties, to fit pieces into an unfinished jigsaw puzzle. But all ideas remain provisional. A few pieces are always missing.

Some people, however, mistake provisional as untrustworthy. Humans' contribution to global warming, for example, is sometimes portrayed as shaky, even though climatologists consider these grounds to be firm. In 2001, the IPCC concluded that most heating of the Earth's surface since 1950 was "likely" due to human greenhouse gas emissions.

And with mounting evidence, this conclusion has become only firmer. Understanding climate change has improved with a greater variety of measurements with wider coverage. By 2007, the IPCC upped its wording about human-caused climate change to "very likely"—an increase in confidence from 66-90% to more than 90%. Six years later, it was deemed "extremely likely" that humans have caused most of the observed increase in global average surface temperature.[4]

What to do? Risk is the combination of probability and consequence. Part of the challenge, therefore, is to safeguard against worst-case outcomes. With business-as-usual, for example, the IPCC estimates a 5–17% chance that temperatures will rise more than 6.4°C by the end of the century—with the spectre of heat waves, droughts, sea level rise, and species extinctions. This probability is well above the point where people typically buy insurance.

At the same time, we are rewriting the odds,

increasing the chances of dangerous climate change.[5,6]

To some, uncertainty is cause for inaction. But public policy cannot wait. We need to invest in the future. Decisions whether or not to immunize against disease, protect wildlife habitat, or regulate greenhouse gases need to be made. Scientists can help lay out the state-of-knowledge and possible consequences; policymakers and the public might use prudence where outcomes look large and irreversible. After all, whether or not you accept the theory of gravity, it's wise to deploy a parachute as a precaution.

We gladly accept risks and rewards as part—indeed the spice—of life. Confronting uncertainty could mean more favourable prospects for tomorrow.

Nature's Alphabet

The Tale of Lord and Lady, Minus a Few Characters

"We're going to deny this project, *because of a duck?*" My buddy cut to the heart of the matter—and it was a mismatch. In one corner was the Harlequin Duck, the elegant bird of fast-flowing streams, affectionately known as Lord-and-Lady. In the other corner, the massive Voisey's Bay nickel find, with its promise of jobs and better days for Newfoundland & Labrador. To my friend, the contrast was ludicrous; the conclusion, obvious. Proceed with the $4-billion development even if a few endangered ducks were lost.

The Harlequin may be a rare bird, but this kind of encounter is not. One by one, nickel-and-duck contests are playing out across the continent—in northern Alberta, where woodland caribou are losing ground to oil sands exploitation; in Montana, where migrating pronghorn antelope are being squeezed by energy projects; in Ontario, where Massasauga rattlesnakes face threats from roads and our misplaced fears. While the benefits of roads and resources are clear, we need to look at both sides of this balance sheet. What does it mean to lose species?

We have an instinctive connection to wildlife. Indeed, the loss of a creature like the Harlequin Duck cuts deeply. We admire the Harlequin for its striking plumage, regal stance, and unique feeding style. Plunging to the bottom of turbulent rivers, this bird walks against the current, picking insect larvae off the stream bed. Its diet includes immature blackflies. (Who wouldn't celebrate that?) Just a glimpse of Lord-and-Lady stays etched in our memory. At the same time, we recognize the bitter finality of extinction. It is the only environmental change that cannot be undone–ever.

Species are also actors in a larger ecological theatre. To witness this grand play, we must consider ecosystems–the intricate, tangled connections among living things and their surroundings. This is no simple task. Ecosystems are not just more complex than we think; they are more complex than we <u>can</u> think.

To get a grip on complexity, we can turn to models. Extinction, for example, has been likened to rivets popping out of an airplane wing.[1] Glancing out the window at 30,000 feet to find a few rivets missing–this may be no cause for alarm. As more rivets become unstuck during your flight, however, your composure could vanish. The breaking point may be hard to predict, but if rivets keep popping, your fall from the sky becomes a sure thing.

Species have been compared to letters, too. Deleting a few characters from the alphabet might cramp your writing style, but is it really serious? Consider this school age predicament (inspired by a true story). Imagine typing an English essay, but with a damaged keyboard: No letter 'E'. You must improvise, and so you resort to the next best thing–

the letter 'I'. With the deadline approaching, straining to stay true to Shakespeare, you type:

"O Romio Romio! whirifori art thou Romio?"

Will your teacher decipher Juliet's celebrated line? Perhaps not. And if you unwittingly disabled more letters, the meaning could be lost or misinterpreted, as in '%#!@' or some other four-letter word. You fail. An unwelcome letter, 'F', appears on your report card.

You should have done better—by replacing the keyboard, of course, or just by typing <ALT>-69 for 'E'. Indeed, substitution is the basis of the modern economy. Over the past two centuries, for instance, North American society has shifted from burning wood to coal to oil and natural gas. And with a petroleum pinch on the horizon, we have faith in the next energy alternative. But not everything is replaceable. As Thomas Homer Dixon cautions, "there are no good substitutes for some of the most precious things nature gives us, like biodiversity and a benign climate."[2] Keeping wildlife isn't just a matter of ethics. It is the foundation of a vibrant economy. The wealth of nations is founded on species and the services they provide, like the supply of food, recreation, clean air and water, and control of pests. And if we conserve nature, this steady stream of benefits will continue to support us, for free.

To an automaton, writing is a collection of lifeless inkblots on paper, or mere pixels on a screen. But as every reader knows, when these simple objects are combined in exquisite ways, higher-order qualities emerge—like imagery and beauty—the very features we deem human. Skilfully joined, inkblots can inform us, enrage us, delight and enlighten us.

And this is where computer keyboard meets Harlequin Duck. If the letters of the alphabet are species, then literature represents ecosystems–the natural environment where we find wonder, relaxation, and peace. Like a good read, the wild expands the mind. To biologist David Janzen, nature is "by far the most diverse and evocative intellectual stimulation known to humans." TV and Xbox are no match for the out-of-doors.

But an extinct species is a letter gone for good–a spelling lesson we haven't fully learned. When Alfred Russell Wallace, greatest field biologist of his day, addressed the Royal Geographic Society more than 150 years ago, he cautioned against extinction, the telltale sign of "a people so immersed in the pursuit of wealth as to be blind to higher considerations."[3] Yet we continue to discard characters from life's alphabet at a disturbing pace. Today, species are vanishing 100 to 1000 times faster than the ancient, background rate.[4] Sustainable, this is not.

Could we risk losing a letter or two? As in "tyre" or "tire", an organism might serve as substitute for the role of another. But the characters of nature are interdependent. Extinguishing a species may cause unforeseen losses–a cascade of secondary extinctions. When a keystone is lost, whole chunks of the alphabet may tumble. Lyrics are impoverished, perhaps irrevocably.

At Voisey's Bay, it is simple to add up the

economic benefits of the mine and mill, but what about the Lord-and-Lady? Maybe in this instance we were fortunate. Shortly after the mine was approved, the species was down-listed from endangered to more favourable status. Better survey techniques revealed more Harlequins than we thought. And perhaps habitat will be restored once the nickel is gone, but it will be decades before we know for sure.

The future of life is unfolding. Our values will be recorded into this narrative, as species lost or conserved, our decisions plain to all future generations. We need a happy ending to this tale— because how we dismantled nature's alphabet is a bedtime story we'd prefer not to tell our grandchildren.

Moon Shot or Pie in the Sky?

March break, for many Canadians, is an annual opportunity to flee south in search of warmth and a little relaxation. But whether it's Disneyland or Daytona Beach, your destination looks humdrum next to Newt Gingrich's travel plans. His vision? Establish a permanent colony on the moon by 2020, with as many as 13,000 residents. To the Republican, the moon is the next frontier, an opportunity for mineral wealth and space observation. Mr. Gingrich even mused about making the lunar colony the 51st US state. The State of the Union address would never be the same.

Comedians were quick to pounce. Stephen Colbert poked fun at the idea of the moon as industrial centre: "America will bring manufacturing to the moon. Ohio? Out of luck." But others see serious potential in a moon colony. China sent its lunar rover in 2013, with possible human settlement by 2025. And to some, extra-terrestrial conquest is critical for humanity. The moon could be a foothold on the future.

Is it stellar vision—or sheer lunacy? For guidance, we can turn to a real, down-to-earth experiment. Enter Biosphere 2—a $200-million prototype space colony built in the 1980s.[1,2] Rising out of the Arizona desert, Biosphere 2 appears as huge, futuristic domes of glass and steel. The goal of the experiment was simple: Design a sealed facility to support 8 people for 2 years with food, water, and air—no materials entering, none leaving. The crew was carefully selected: 4 women, 4 men, ages 29 to 67, with diverse skills including diving, gardening, plumbing, and scientific observation. Indeed, in psychological testing, these "Biospherians" scored better than astronaut candidates.

But having the Right Stuff wasn't enough. To survive 24 months, they needed the Right Species—nearly 4,000 kinds of plants, fungi, and animals that would serve as sources of food, oxygen, clean water, and other essentials. This was Survivor, well before reality TV became commonplace.

As it turned out, life under glass was complex. Social problems were few, apart from a violent incident where one crew member threw a teacup at another. But some challenges were unanticipated. About 1½ years into the mission, levels of oxygen fell—barely adequate to keep the Biospherians alive—while carbon dioxide levels soared. The culprit was unforeseen microbial activity in the soil. Vines, like morning glory, were expected to help store carbon. But they needed pruning to keep them from overrunning other vegetation, including food plants. Tree trunks and branches turned brittle and were liable to collapse. Water became polluted; the atmosphere was poisoned by nitrous oxide. Experimenters had to intervene.

And then there were losses. While some extinctions were expected, whole groups of creatures disappeared. All pollinators went extinct, cutting short the life cycle of plants that depended on them.

Most insects vanished, leaving a hoard of crazy ants, one kind of grasshopper, and a few cockroaches. Food was scarcer than expected. The Biospherians fell short of their goal of 2500 calories per day. Despite large energy subsidies and brave individual efforts, the cycles of nature could not be made to work.[1,2]

This outcome is humbling. Indeed, while some peg our future on technological solutions, ecology suggests otherwise. To date, no one knows how to duplicate the life-support services that nature provides for free. And ecosystems, we have learned, are more complex than we can think. For the foreseeable future, the prospects for humanity are likely to be shared—with nature. Biosphere 2 is a cautionary tale.

But where, you ask, is Biosphere One? Well, that is Earth—the only home that you and I shall ever know. And, for the sake of our species and countless others, we'd better look after it.

LIFE IS A HIGHWAY

Driving is mundane, rarely deserving a second thought. But here's a curiosity: Why drive on the right? Naturally, you want to abide by the law and avoid a head-on collision. Your behaviour may more ingrained than you appreciate. If you learned to drive on the right, then I suspect you unconsciously gravitate that way–even while strolling a busy sidewalk or manoeuvring a shopping cart down the aisle. Right just feels right.

A trip abroad can reveal the depth of your leanings. Visit Britain. To your unaccustomed North American eye, roadways and automobiles appear like strange, mirror images. Traffic approaches from the wrong direction; the passenger is driving. As a pedestrian, you are warned to LOOK RIGHT before crossing. With good reason: Research shows visitors from right-hand-driving countries are more liable to be struck by vehicles.[1]

Around the world, right-hand driving predominates. But in nearly one-third of countries, motorists keep left–in the British Isles, New Zealand, Australia, Japan, plus much of southern Africa, Indian subcontinent, Caribbean, and Micronesia. Some historians trace the left-right distinction to an age-old bias, a tendency to free one hand for a horsewhip or weapon. Once established, laws and custom spread in the wake of Roman, Napoleonic, and British conquests.

Left- or right-hand driving, therefore, may be

little more than historical accident. What's crucial is conformity, everyone keeping to the same side—even across the border. Around the world, countries tend to mimic their next-door neighbours. Left-handed driving also tends to occur on islands, countries less constrained by behaviour over the border.[2]

It wasn't always this way. In the early 1920s, Canada was a patchwork. Left-handed Loyalists ruled the roads in British Columbia and the Maritimes. But safety and simplicity demanded uniformity. All provinces adopted right-handed driving by 1924. Even Newfoundland—ever the distinct society—changed in 1947, just prior to joining Confederation.

Changeovers are rare. And no wonder, given the confusion, costs, and inconvenience of replacing road signs, dealing with awkward steering and bus doors, and dispensing with die-hard habits. The Pacific island of Samoa managed a wholesale flip—the first in decades—from right- to left-hand driving at 5:30 AM, Monday, September 7, 2009. According to *The Guardian*, the change was accompanied by a 2-day national holiday to lessen traffic, and a 3-day ban on alcohol sales to reduce accidents. No incidents were reported.[3]

Favouring one side is what scientists call lateralization—bias toward the right or left. In people, lateralization is so pronounced (90% of us are right-handed, 80% right-footed) that you might regard it as uniquely human. At one time, so did scientists. But as evidence grew, researchers realized that many animals, from insects to primates, show lateral bias.

The list includes chimpanzees, our closest relatives, who tend to be right-handed when using tools and throwing objects.[4] On the other hand (pardon the pun), ruminating cows are prone to lying on the left.[5] And remarkably, some species of bumblebees approach flowers in a consistent way, circling either clockwise or anti-clockwise.[6] Asymmetrical in behaviour has been reported in more than two-thirds of species.[7] Nature is more handed, footed, and pawed than we supposed.

And like motorists, right-left bias in animals seems to have a social cause—the need to coordinate activities of the group. Among fishes, for example, one study found that all schooling species displayed a turning bias; fewer than half of non-schooling species did so.[6] To socialize is to move in harmony. It's a link between the modernity of the roadway and the ancient order of Nature.

Understanding the animal world is more than a curiosity. It gives us a deeper sense of our place in it. For social creatures, life is a highway.

Expect Delays, in Slow Motion

With each election campaign, political leaders seem take aim at one healthcare issue: Wait times, those troublesome lags between referral and medical procedure. But beyond the doctors' offices and hospitals, another kind of delay is affecting lives and refusing to budge. And these stubborn lags are not measured in days or months—but decades.

Such is the tempo of the environment. It is the other big healthcare provider, the source of vital resources like clean air and water. But unlike our hectic lives, running on fast-forward, ecosystems often move at a leisurely pace where even serious shifts can be subtle. Because while it's a snap to notice abrupt changes, we may fail to grasp events that unfold over decades or centuries. And we need to prepare for the impact of these slow-motion crises.

Consider climate change—the hidden emergency that has moved to the forefront of public concern. Atmospheric scientists tell us that most of the heating of the earth's surface since 1950 is attributable to people, especially to land use changes and fossil-fuel CO_2 emissions. And the prospects are for further warming, perhaps as much as 4°C this century. With it, we can expect a host of undesirables, like extreme weather events, sea level rise, and species extinctions.[1]

But science also informs us that the full impact of today's atmosphere will take decades to be realized. Even if carbon emissions were instantly cut to zero, warming is likely to continue—the consequence of the inertia of water and sky. CO_2 is a long-lived greenhouse gas, persisting in the atmosphere for decades or more; and ice sheets and oceans are slow to absorb the added heat. This environmental delay connects us in unintended ways. Generation by generation, we inherit the climate change caused by previous CO_2 emissions, especially from our parents—and we bequeath the impacts of current emissions to our children.[2]

Belated effects are also important for wildlife. Take woodland caribou, creatures of the boreal forest. In northern Ontario, human encroachment has caused widespread caribou decline, but disturbing the forest does not extinguish the species outright. It sets in motion a complex series of habitat changes: a flush of new vegetation which attracts moose, followed by more predators and the demise of caribou. It's a slow falling of dominoes, taking some 20 years[3]—a period that may outlast the careers of many present-day professionals.

In the human world, of course, postponements can be valuable as opportunities to reconsider and readjust. A few years ago, the CBC brass thought it useful to impose a 7-second broadcast delay on hockey guru Don Cherry, his penalty for an on-air rant. And in the classroom, teachers typically afford students a mere two seconds to respond to questions. Giving kids a little more time to reflect can improve their learning.

But, as author Elizabeth Kolbert points out,

environmental delays are two-sided.[4] On one hand, they grant us space to get ready, to rescue endangered species or make our communities more resilient. Preparations can be as simple as planting trees. In California, it is estimated that strategically adding 50 million shade trees, to relieve the strain on air conditioners, would provide energy savings equivalent to seven 100 MW power plants.[5] Foresight is embodied in the act of planting a tree.

On the other hand, delays are a potential hazard, waiting to ensnare the unsuspecting. They may cloud our ability to read the true state-of-affairs. More insidiously, they could represent false licence for business-as-usual–leaving the costs of disrupted climate and diminished biodiversity to our children and grandchildren.

What to do? We could resort to fatalism. Or, as the only species that knows it lives on a planet, we could regard this knowledge as a call to wisdom–an invitation to deal now with the expected consequences of the future. To Stanley Davis, author of *Future Perfect*, this is the stuff of great leaders. Success entails managing "the beforemath", the outcome of events that have not yet taken place, rather than picking up the pieces later on.[6]

Our environmental future calls for anticipatory actions–the combination of promptness and foresight. To do so would make a leader of every one of us.

A Brave New World for Birds and Beasts

Is location a thing of the past? Uprisings in the Middle East hint at an emerging, boundless world. From unlikely beginnings in a Tunisian market, the demands for democracy rocked governments from Libya to Bahrain–but they also revealed the power of social media to cut down borders. The video tablet now seems mightier than the pen. Even internet names, with domains like .com and .org, suggest a world where location no longer matters, where ideas respect no bounds, where information carries no return address. Geography looks passé.

Not so fast. Despite the electronic swiftness, there is no denying the importance of locale–the neighbourhood and community where we grow up, work, and live. People are the product of their surroundings. The evidence is at our doorstep: The battles to keep local schools, the ups and downs of the real estate market, NIMBY, even simple acts of neighbourly kindness. Politicians know the importance of reading the landscape, too. "Everything has to do with geography", said Montana Governor Judy Martz.

There is still a place for place. And when we add living things to this mix, geography acquires special meaning. The whereabouts of plants and animals conveys more than mere curiosities or brute facts. It reveals the secrets of the past–and offers a roadmap for the future.

A Brave New World for Birds and Beasts - 84

Location, location, location. A modern cliché, but also the foundation for one of the most staggering ideas ever–evolution by natural selection, or how countless species on earth came to be. Charles Darwin began decoding this Mystery of Mysteries while still twenty-something, during a 5-year, round-the-world voyage. As ship's naturalist, Darwin had to sail like a mariner, to think like a detective, and to collect like a man possessed. But he also needed to be an astute reader of geographic clues.

The Galapagos Islands proved a landmark on this journey.[1] Darwin was fascinated by these islands, often within sight of each other, and inhabited by somewhat different creatures. Like connoisseurs, local residents could distinguish a tortoise from one island or another by just slight variations–differences in size, the shape of the shell, even subtleties in taste. Finches, mocking-thrushes, and plants also varied from island to island. To Darwin, this was unexpected. Isolation, he realized, sets the stage for new species. When separated, living things adapt to local conditions and slowly diverge into distinct forms. And geography is key to the diversity of human languages and cultures, too. Take the Canadian city of Peterborough and the Italian city of Venice. The two communities share many features–similar latitude, appreciation of fine food, and love of waterways. But there are obvious differences. (Indeed, while gondolas may be rare on the Lift Lock, skating is ill-advised on the Venetian Canalasso.) *Vive la differenza!*

Geography infuses variety. It is the reason we travel, explore, take photos—and without it, there would be less diversity across the living world, too.

But the ancient arrangement of living things is being shuffled. Sometimes accidentally, other times intentionally, the biological uniqueness of place is being washed away. And this erosion stems from two powerful, modern forces: Cities, which homogenize the environment,[2] and transportation, which inadvertently carries plants and animals outside their home and native lands.[3-6]

We know how uniform cities can be. Stroll through any shopping mall—and you may be hard pressed to discern whether you are in Peterborough, St. Petersburg, or Penticton. *Go, Petes, go ... anyone?*

But what's unexpected is the biological sameness of the city. Consider a trip unimaginable to Darwin—halfway around the world in a day-and-a-half. A flight from Quebec City to Hobart, Tasmania, would whisk you across 15 time zones in just 35 hours. When you arrive, plagued by jet lag, you may have trouble finding your bearings, not the least from the birds outside the hotel window—you see starlings, house sparrows, pigeons. You are startled to find such common feathered friends in a far-flung place. These birds, like so many brands of clothing and fast food, are recent features, spread to places with no historic connection. Quebec and Tasmania have not been linked by a land bridge for some 200 million years. But the sameness of cities provides uniform habitat—and we moved these creatures across oceans and continents to suit our tastes, in the geological blink of an eye.[2]

Although half-a-world distant, Quebec and Hobart are more similar in bird life than two nature reserves separated by a day's drive. Consider another unlikely pair: Boston and St. Louis, nearly 2,000 km apart, have more plants in common than two national parks within walking distance of one another.[2] Biologically, we've pulled the world together.

Don't we crave sameness? Sure, but there is a downside—the fate of native species. Exotics, without predators or other usual limits, may explode in numbers and displace indigenous plants and animals. Back in Peterborough, add European starlings, we lose bluebirds; introduce zebra mussels, we lose many native mussels; bring in Asian long-horned beetles, we risk devastating maple forests. While a handful of alien species prosper, many native species face extinction. It is estimated that half the world's mammals would be lost if barriers were dismantled. This is The Great Levelling—a homogenized world of few biological winners and countless losers. And we lose, too. Biodiversity is key to life-support systems for people, like the continual flows of freshwater and food.

But there is good news—straightforward ways to check the spread of exotics. Some measures are international, others local. To safeguard lakes and rivers, for instance, ballast water on ships can be exchanged on the high seas rather than into the Great Lakes. Local fishermen can refrain from dumping their bait buckets. To protect forests, shipments from overseas can be inspected and fumigated for pests. And on your next camping trip, you can leave the firewood at home—along with the unwanted insects it might harbour.

We have a deep, eternal connection to living things. Nature affirms our sense of place, lends timelessness to our lives, and connects us across generations. To uphold this trust, we need to chart a course for the century—a journey that will be conceptual as much as geographic. "One's destination is never a place but rather a new way of looking at things", said author Henry Miller. Keeping a little space for nature, within ourselves, would be a good point of departure.

The Most Successful Enterprise

Good Advice Is a Science

Here's some sensible advice: Be careful about giving advice. That's what some scientists must be thinking after six of them landed in jail, convicted of manslaughter. Just days before a deadly earthquake in the Italian city of L'Aquila, seismologists gave reassuring but misleading public statements about the dangers of tremors being felt by local residents. Then, a 6.3 magnitude earthquake killed more than 300 people. A judge handed the scientists 6-year sentences for giving inaccurate information.[1]

The convictions sent shockwaves through the scientific community. Indeed, researchers are regularly asked for their views on the hazards of earthquakes, contaminants, climate change, and spread of disease. And prediction is one of the values of science. For instance, biologists have determined that knowing the body length of a fish–a simple measure–can be used to predict the size of its home range, the area needed to meet its life requirements.[2] This is useful, given that space is crucial for wildlife to thrive.

One of the most celebrated predictions comes from one of the greatest scientists. In 1862, Charles Darwin came across an astounding orchid in Madagascar–a flower with a remarkable, foot-long tube that held nectar at the tip. Darwin understood the intimacy between plant and pollinator. From theory, he predicted the existence of an unlikely insect, one with a similarly long feeding tube or proboscis. Biologists were skeptical. They had never encountered such an insect on the island.[3]

Decades after Darwin's death, a discovery: A hawk moth, capable of hovering like a hummingbird and outfitted with an impossibly long, lash-like proboscis. It was given the scientific name, *Xanthopan morganii praedicta*, in recognition of Darwin's foresight. While this is famous case, predictions—generated by theory and tested by experiment—are routine for scientists.

Of course, forecasting is not exclusive to science. Day traders and sports enthusiasts often seek expert opinion in hope of improving their odds. But such forecasts, it seems, are seldom tested. I decided to examine their accuracy. I bought an aged copy of the 2004 Hockey Scouting Report (at 25¢, a cheap find) and checked its NHL player predictions against actual performance. The test—to see if expert opinion could beat naive expectation, based on a player's performance the previous year.

How good were the experts? While their forecast was better than random, it was no more accurate than relying on past performance.[4] What's more, experts were overly optimistic at the low end of the scale, overestimating a player's performance on average by 13 points.

Of course, hockey stats are trivial next to an

earthquake's devastation. And therein lies the difference. The L'Aquila tragedy was not just about prediction—it was about risk, the combination of probability and consequence.[5] And there are reasons to err on the side of caution. The smoke detector in your house, when you burn the toast, may be a nuisance, but you hesitate to disarm it. Better an alarm without fire than fire without an alarm.[6]

The more unpredictable the world, said author Steve Rivkin, the more we rely on predictions. But accurate forecasting remains difficult in some areas like pandemics, meteorology—and the timing of earthquakes. The Italian seismologists, it seems, did not clearly communicate the risks. It was a quake with the alarm disabled.

Canadians trust deeply in science to improve their quality of life. In return, scientists have a key role—to convey clearly and honestly the outcomes of choices facing the public. In the aftermath of L'Aquila, one of the causalities might be scientific dialogue, as researchers shy away from public engagement. No advice would be an unhappy outcome for all.

THE MEASURE OF THINGS

It's time to fess up: Your science teachers didn't tell you the whole story. In textbooks and lessons, we served up the facts and figures and declared that science doesn't just seek answers; it seeks <u>the</u> answer. But we have an admission to make.

Confessions can be awkward, so let's deal with this in a scholastic way—with a quiz. Here's a little test that leads to some surprising answers. Our first question is from geography.

Question 1. How long is the coastline of the island of Newfoundland?
(A) 11,548 km
(B) 9,656 km
(C) 2,825 km
(D) All of the above.

Even without resorting to an atlas or *Geography for Dummies*, surely we can eliminate "all of the above." A coastline, sturdy as bedrock, has only one length. Or does it?

Imagine yourself at the Newfoundland shore—rugged and indented with many bays and coves. To measure its length, you might trek the entire coast, counting each stride as you go.

But suppose you shrunk to the size of an ant. A diminutive creature, you must now negotiate each rock and pool, and the coastline would seem much longer. Or, transformed into a giant, you could make short work of it, straddling peninsulas and skipping over coves. Human, ant, and giant would experience the very same coastline in dissimilar ways, and each would cover much different distances.

In everyday life, we are accustomed to one question = one answer. But science has discovered that changing the window size on the world–different scales–can lead to different conclusions, each correct, yet none by itself providing the whole picture.[1]

But all scales are not created equal. Many activities in nature are linked and happening at the same time. Theory tells us big-scale processes may be more important. Slower, they set limits on small-scale processes while, in return, small activities explain the big ones.[1,2]

Take a familiar event–buying a house, a rare occasion in one's life. Once you've settled in to your home, choosing a room to browse the evening paper occurs much more often. Your choice of interior reading space is constrained by the house you've purchased, while those very rooms–their size and ambience–explain why you bought the home in the first place. Big things constrain; little things explain.

It's an idea with practical implications. Consider alien species–organisms that have spread beyond their native land. In Ontario, aliens are all around us: dandelions, starlings, house mice, and countless other plants and animals. We've arrived at the next question.

Question 2. Which statement is correct?
(A) Alien species increase biodiversity; sea lamprey, zebra mussels, and round goby would otherwise not exist in the Great Lakes.

(B) Aliens decrease biodiversity; they cause extinctions by eating or out-competing native species.

Another unexpected answer: Both A and B are true, but the big-scale process, B, is more important. While exotic organisms can generate local or fleeting gains in biodiversity, they drive extinctions–the irreversible loss of species worldwide.[3]

Bigger, slower, more significant. This is a principle we might use to sort out the pivotal issues of the day. Assemble a list of top environmental concerns to society: climate change, cross-border contaminants, collapse of migratory stocks, spread of exotic species and diseases. Each item stretches to continental, oceanic, or global proportions.

Our institutions are ill-equipped to deal with such issues. Governments and corporations react smartly to acute events, with a typical focus on the next fiscal year or next election. It's a scale mismatch.[4] Concentrating on here and now, we are inclined to relegate big problems to another place or time.

Size matters. But what about private incentives and local circumstances? How can we unite interests, personal to planetary? One possibility is full-cost accounting of our actions, an economy where conservation represents financial gain to individuals as well as society. Objectives, large and small, become aligned.

And that's the final question, still unanswered, certain to dominate the 21st century–whether we will adjust to the biggest constraint of all, earth's biological support system.[5] Scale helps frame the issue. Zoom out and we comprehend the planet at its fullest—no longer a limitless well or bottomless wastebin as once supposed, but as the only place where all life unfolds.

Scale is part and parcel of the most fascinating questions we face, in the classroom and beyond. The answers are not merely academic; they will shape our future.

EVIDENCE MEANS A SPORTING CHANCE FOR ALL

It's in the news, each and every day. It compels the attention of royals; it can lift the disadvantaged to fame and fortune. Thousands are moved to cheer, declare their allegiance, and gather in the streets.

This is modern sport. And whether you are a fan, a participant (or neither), you appreciate how sport can grip the nation. Indeed, with just a few words—"Sidney Crosby! The golden goal!"—you can conjure up the image, the elation, the maple leaf.

Sport has become an unstoppable social and economic force. Three out of four Canadians take part as players, coaches, or spectators. It boasts the second-highest proportion of volunteers in the country.[1] And revenue from spectator sports is rising, even as leisure time declines. In our atomized world—everyone with a personal screen—sport remains a shared experience. No wonder Rogers Communications signed a $5.2 billion deal to broadcast NHL hockey.

But the foundations of the game, the reason we continue to watch and play, are rooted in rules and reason. We insist on the utmost integrity of officials; we demand the best available evidence to determine a result. And in televised sport, truth is laid bare for all to see. A disputed goal can be replayed in exquisite, slow-motion detail -- sometimes leaving the referee as the lone soul who missed it. To disregard evidence is to invite outrage. *We were robbed!*

Evidence Means a Sporting Chance for All - 98

Technology has changed the game. The old ways of officiating–snap decisions with the unaided eye–have collapsed under the weight of video evidence. The CFL and NHL now regularly use frame-by-frame review, from a dozen or more cameras, to help resolve contentious calls.

Other sports have been slow to adapt. But even FIFA, soccer's ultra-conservative governing body, has introduced goal-line technology to the World Cup. Seven high-speed cameras focus on each goal and triangulate the exact position of the ball. Once a goal is scored, the result is relayed to a watch worn by each official. The delay? One second.

Evidence, evidence, evidence. Getting it right is crucial, but good decision-making is not exclusive to sport. Away from the stadium, we often grapple with choices about health and wealth. Delivering the data is the role of science.

And here, too, new tools can improve our powers of detection. Consider a long-contentious issue: seeping toxins from the Alberta oil sands. In a recent study, federal researchers used advanced technology to detect a telltale chemical signature, distinct from natural sources. The evidence confirms chemicals are leaching from tailings ponds into groundwater and Athabasca River.[2]

Sporting judgements can be quick; environmental assessments may take much longer. Look at the Great Lakes, source of drinking water for 24 million people. Since the 1970s, scientists have been monitoring the eggs of herring gulls. Because these waterbirds are year-round residents at the top of the food chain, they are sensitive indicators of contaminants—an exceptional camera angle, if you will. And the data are encouraging. Most pesticide-related compounds, like those from DDT, have been declining.[3] From the perspective of decades, we can make this call with confidence.

Of course, science does not dictate decisions or prescribe policy. But to blithely disregard evidence—worse, to halt monitoring, to cut off the data flow—could be perilous. This is why researchers were dumbfounded by the snap federal decision to close the Experimental Lakes Area, a unique vantage point on freshwater quality.

Sport can be revealing. And it shows us as a fair, determined, and enlightened people—a science nation. In the long run, evidence will be key to our game plan.

OF MICROBES AND MEN

You will find them, living quietly among us, in virtually every community–the lucky ones. They are the people who owe their lives to some heroic act or good fortune. Their stories are gripping: the baby pulled from rubble of an earthquake; the distressed swimmer rescued from icy waters; the heart attack victim resuscitated by CPR. Their true-life tales make headlines, sometimes even heroes.

I am one of these lucky individuals. While not something that preoccupies me, my existence would have been cut short without some fortunate circumstances of time and place. My saviour was less dramatic, but no less life-saving: Penicillin. Wonder drug, it delivered me from deadly grip of childhood pneumonia. And this disease remains the number one killer of children–more lethal than malaria, measles, and AIDS combined.

Human history is marred by the scourge of infection–pneumonia, tuberculosis, typhoid, the plague–diseases that levelled armies, societies, and empires. Prior to the mid-20th century, we inhabited a world where the chance of untimely death from contagious disease was as much as 40%. Succumbing to infection after trauma or childbirth was not uncommon.

Antibiotics rewrote these odds. Many experts view them as the most important medical breakthrough, ever. In the late 1960s, so confident was the US Surgeon General that he pronounced it "time to close the book on infectious diseases and declare the war against pestilence won." In the clash between humanity and microbes, we had prevailed.

But bacteria proved formidable foes. It wasn't long before resistant strains appeared. Witness the relentless march of stubborn bacteria: MSRA, CRE, VRE, and MDR-TB. Their common name is R for resistant. Some, like *Salmonella typhi*, can defy first-line, second-line, even third-line antibiotics. Resistance is growing faster than new anti-infective drugs can be devised. Our antibiotic arsenal is slipping.[1]

We might have seen this coming. Antimicrobial resistance results from a natural biological process—evolution. And bacteria, because of their great numbers and speed of reproduction, have some potent advantages. Their generation time is as short as 15 minutes. Evolution acts quickly.

In hindsight, perhaps "the war on microbes" was the wrong analogy. These creatures outnumber us by more than a billion-billion-fold and reproduce 500,000 times faster. To avoid sliding back into the pre-antibiotic era, we must play to our strengths—our large brains. In the words of Nobel laureate, Joshua Lederberg, it is "our wits versus their genes".[1]

Knowledge is our ally. And it discloses the reason for resistance. Imagine the bacteria inhabiting a patient treated with antibiotic. Like any population of living things, these bacteria differ genetically. Some will be better at withstanding the tide of medication. Defenceless bacteria are eliminated; the survivors reproduce, passing the gene for resistance to the next generation. Resistant bacteria become the common type, potentially transmitted to other patients.[2]

This is the crux of Darwin's theory. And while evolution tells us antibiotic resistance is inevitable, it highlights some simple measures to slow its pace—like judicious use. Indeed, antibiotics are ineffective at combatting viral infections like colds and the flu.

Evolution is often regarded as an ancient enterprise, unfolding in the long sweep of geological time. But it is also the hidden hand in contemporary issues—why DDT fails to eliminate malaria-carrying mosquitoes, why trophy hunting leaves fewer trophy animals, why commercial fishing reduces the size of fish.[3-5] Evolution reveals why: Removing animals simply selects for more elusive individuals, leaving smaller or resistant ones to reproduce.

Something old, something new. Evolutionary theory is a venerable idea; it is also our guide in modern matters of health and wealth. Although he didn't know it, Darwin is a hero in this 21st-century parable of medicine and microbes.

Translation Is a Tricky Business

Long before the printed word, eons ahead of Twitter, our human ancestors acquired the most human of all human traits—language. Some scholars regard language as our secret weapon, the instrument that led to the rise of humanity and our domination of the planet.[1] Of course, other animals have astounding means of communicating—bird song, mammal scent, the bee waggle dance—but nothing in nature rivals human language for its precision, abstraction, and emotion.

And since the dawn of language, some 100,000 years ago, the diversity of speech has exploded. Today, there are nearly 7,000 languages spoken around the world. But linguistic diversity brings linguistic divides. Sometimes we need to speak in, or to listen to, an unfamiliar tongue.

Translation is a tricky business. Being bilingual is no guarantee, I've learned, that you can interpret phrases crisply and cleanly. For example, my wife speaks rudimentary French. After travelling in France for a few weeks, she learned *composter*—to stamp—a train ticket. Funny, even now, this word is lodged in her vocabulary; she cannot readily find the equivalent in English. On the local GO Train, we *compostons* our tickets.

Modern translation tools are helpful, but they're not without perils. To illustrate, I ran our national anthem through Google Translate—10 different languages, such as Chinese, French, and Yiddish, each in sequence—then translated the lyrics back into English. Here are the results, the first six lines of O Canada [with my annotations]:

Canada, Oh! Native and our home [Good start]
Patriotic love true secondary control of all files [Clear reference to Harper government]
Please see the shining climbing centre [Obviously the CN Tower]
The true north strong free
Of course, throughout Canada
We stand on guard for you. [Better than the original!]

The message can become garbled. And linguistic boundaries can be more subtle, too, lurking in hidden places. English, for example, is the universal language of science. But the truth is that we scientists adore jargon—our shorthand, our insider code, our way of reaffirming ourselves as specialists. Jargon is a brick wall, however, when we speak to fellow citizens.[2,3] And speak we must. After all, it is the Canadian people who fund science and who trust in science to enhance their quality of life.

I regularly challenge my students to find alternative words, in plain English, for jargon—RNA, for example. They are often stumped, unable to suggest anything beyond "ribonucleic acid". Oddly, their learned minds seem unable to revert to simple language. This is The Curse of Knowledge.[4] By acquiring what they now know, they cannot recall what it was like when they did not know it. Established scientists are even more severely afflicted. The curse is difficult to shake.

Yet shake it we must. Scientists need to convey their knowledge, in accessible language, to help guide the public with the most pressing issues we've ever faced–the looming threats of climate change, antibiotic resistance, mass extinction, infectious diseases.[5,6] I have never met a scientist who doesn't want to see his or her results broadly known, widely discussed, and applied to such problems. But the first step is bridging the communication gap, from scientist to non-scientist.

There is good news. More and more, my students are interested not solely in science, but in learning how to communicate it effectively. The next generation of scientists is the new and improved Google Translate–brimming with knowledge and adept at conveying it. They have arrived just in time.

LOOKING PAST THE COIN FLIP

Kicking off a new academic year is a time for renewal–and confessions. This year, the unspeakable happened during my university ecology class, some 200 students, the lecture hall abuzz with anticipation that signals the start of term. It's here, the first class, where I like to mobilize students' enthusiasm: Not merely to lay out deadlines, but to instill excitement for ecology, my lifelong passion.

What was this heresy? The students and I strayed beyond the usual bounds, perhaps into pseudoscience. We conducted an experiment, not about plants, animals or typical ecological stuff, but telepathy.

And this was not my first offence. Last year, too, I carried out the experiment with students. At the front of the class, I announced I would flip a coin four times. Each time, without disclosing the result, I broadcast a telepathic message into the room–whether the flip was heads (H) or tails (T). I asked students to write down what message they sensed.[1]

The result was then revealed: T T H T. Who, I asked, had managed an exact match? More than a dozen students raised their hands. *Yes!* Consider the implications, I proclaimed. Rather than speaking audibly in lecture, I'd simply meditate the key concepts. And during exams, students could ditch pens and pencils, and merely think their answers.

But wait. When asked who had recorded the converse (H H T H), roughly the same number raised their hands. Hmm. Finally, I enquired, how many had written T T T T or H H H H? Not a single student. But weren't these sequences as likely as any other? (Indeed they are.)

Our experiment exposed some important truths about science—and ourselves. To many, science is a body of knowledge and scientists, walking encyclopedias. But as author Michael Shermer noted, science is not a thing but a verb. It's a way of searching for rational explanations about the world.

And skepticism is the tool that helps dissect real from bogus. It is foolish to accept ideas uncritically, without evidence—whether telepathy, crop circles, or the widow with £500,000 inheritance to share; just email your credit card information. The skeptic demands compelling evidence.[2] And when faced with rival explanations, the skeptic favours the simple over the fanciful. Indeed, our experimental results were consistent with sheer luck. Neither my students nor I were convinced of telepathy.

Believing unfounded claims (and divulging your credit card number) can be damaging—to you and others. Consider imperilled wildlife, like sharks and rhinoceros, edging toward extinction because of overharvesting.[3] Rising demand has pushed the price of rhino horn to $65,000 per kilogram—now pricier than cocaine, gold, or diamonds. Poaching has skyrocketed.[4]

This ecological tragedy is propelled by unsubstantiated beliefs about rhino horn as fever remedy, shark cartilage as cancer cure. Hard evidence is scant or absent. Most trials with rhino horn have failed to find anti-inflammatory benefits.[5] The popular claim that sharks don't get cancer is false; no controlled experiment has determined shark cartilage can combat tumours.[6] Far from benign, such beliefs may divert patients from effective, conventional treatments. Skepticism is healthy in more ways than one.

And it is driven by data rather than ideology. Atmospheric scientists report climate change very likely attributable to human carbon emissions. How does that explanation stack up? According to physicist Richard Miller: "A better match than anything else we've tried ... To be considered seriously, an alternative explanation must match the data at least as well." While research should continue, a scientific consensus has formed on the reality of human-caused climate change.[7] The facts point to us.

You don't need to be a scientist to practise skepticism. But it takes practice. My students and I will do the unconscionable next year, too.

A Little Feedback on Your Weight

If you're like me, there are clothes lying dormant in your closet–pants or jeans that once fit oh so well. A few pounds gained, a few items retired from your wardrobe.

Body weight may rachet upwards, but what's surprising is its stability over the long haul. It defies arithmetic. Armed with a calorie counter and exercise chart, you could easily miscalculate by 100 calories a day–just three bites of a hamburger–and your weight would drift erratically. We would be a population of scrawnies and hefties.

The stability of weight lends support to the idea of a 'set-point'–a control mechanism in the brain that regulates body mass within a steady range. Trends support the theory. Since the 1970s, average daily food intake in the US jumped by 168 calories for men and 335 calories for women, while job-related activity dropped by 142 calories. On balance, the typical American should have gained 18 to 68 pounds per year–a big, fat overestimate of what happened. In each of us, scientists reckon, there must be physiological processes maintaining energy balance.[1, 2]

Dieters know this. Losing a pound is more complex than cutting out 3500 calories, as often computed, because the body kicks in to resist the change. Restricting intake produces weight loss, but it triggers other responses, including lower energy output and a rise in hunger. To lose weight is to do battle against one's own physiology. It's an adaptation that benefited our ancestors in the face of famine.

You owe your existence to feedback. And these systems are more widespread than we might expect. They are powerful forces, often hidden from view.

Society depends on them. We duly complete tax returns, write exams, hunt or fish or shop or drive, abiding by the laws of the land even when no one is watching. Never would there be enough police, invigilators, and auditors to enforce these behaviours. Rather, people act in a civic-minded way when most others act in a civic-minded way.[3] They reciprocate. Peace, order, and good government are rooted in feedback.

On the other hand, stubborn forces can also ensnare people—and keep them there. Poverty, for instance, is not just about being poor, but staying poor. Being born or thrust into poverty may engender conditions that perpetuate it.

This is because its effects are not just economic, but psychological and physiological. People that experience a financial shock, such as losing a job, have higher levels of the stress hormone, cortisol. Poverty can alter behaviour. Credit is hard to secure, so the poor tend to choose a safe and immediate income, even if smaller, rather than a larger, future income.[4] To halt this spiral may require access to pension plans or housing, for instance.

Other feedbacks lead to runaway behaviour. Consider an unworldly example: A few years ago, a communications satellite and defunct military satellite collided while moving at more than 10 km per second, 800 km over Siberia. The crash left an orbiting cloud of debris, roughly a thousand fragments of space junk. It has scientists concerned.[5] Each collision increases clutter, increasing the risk of further collisions, more clutter, more collisions (you get the idea), potentially leaving low-earth orbit unusable.

Positive feedback is at work on earth, too. Higher temperatures, for instance, can cause more frequent and intense forest fires and insect outbreaks, fuelling more carbon emissions—and fast-forward acceleration of global warming.[6] Although hard to predict, these are triggers better left unpulled.

The world is rife with feedback loops—positive and negative, vicious and virtuous. Understanding them can help us deal with some weighty issues, at home and beyond.

Auto Pilot and Points North

Build it and They Will Bike

The conflict has been drawn-out and divisive. For decades, battles raged in the streets; citizens have been divided. And these hostilities haven't played out in the Middle East or some remote part of Africa, but much closer to home. The front lines are less than 2 hours from my home city of Peterborough. This is the War on the Car.

Some, like Toronto Mayor Rob Ford, declared the War over. But there is no doubt the car prevailed long ago. For most of a century, we fashioned our communities, day-to-day lives, even our thinking around the motor vehicle. Indeed, when I say "2 hours" from here, do you imagine a walk or airline flight? Any glance at our surroundings confirms the rule of the automobile—the web of roads, acres of free parking, the Drive Thru at banks and coffee shops. Never has it been easier to grab a few dollars or doughnuts. Life now moves at the speed of the accelerator pedal.

Today, Canadians own 1.5 cars per household—a fleet of more than 20 million.[1] If all the cars in the country were lined up bumper-to-bumper, it would be … well, a typical morning commute. And that is the downside of car culture: Not just snarled traffic, but exhaust, urban sprawl, and a nation of expanding waistlines.

But the times are a-changin'. In the US, people are now driving less—billions of kilometres down from the all-time high in 2005. After accounting for population change, Americans' driving habits now resemble those from 1997.[2] The auto clock is rewinding. Some may regard this as an indicator of an ailing economy, but there is an upside.

It is part of a generational shift. When I was 16, my buddies and I longed for our drivers' licences and set of wheels. Now, teenagers need to be coerced into obtaining their G1 and G2. Driving is no longer regarded as a rite of passage into adulthood. Today's youth have started down a different path.

On this new roadway, the bicycle is making tracks. Not merely a child's toy or Sunday afternoon plaything, the bike has matured into a working machine—for running errands, going to work or shopping. In Toronto, home to roughly a million adult cyclists, about one in three uses the bicycle as a utilitarian vehicle.[3] And these numbers are rising. The streets of the future will be a mix of two-wheelers and four.

And drivers will see cyclists as allies. Indeed, bicycles bring loads of advantages to motorists: Eased traffic congestion, more available parking, less noise and smog, decreased wear and tear on roadways, fewer greenhouse gases, and reduced demand for gasoline and health care. If you can never be too rich or too thin, the bike is the new fashion statement: Healthier citizens and fatter wallets for all.

Build it and They Will Bike - 119

But we have a way to go on both fronts. Today, the typical car trip carries just 1.62 persons per vehicle (driver included), slipping from 1.68 occupants a few years ago.[1] Vacant car seats mean more crowded roads. And in the city of Peterborough, only 2.3% of workers cycle to work. Safety is often cited as a concern.

The recipe for security and compatibility is as simple as the bike lane. In Toronto, the mayor has seen the future. He recently announced plans for 80 km of on-street bike paths and 8,000 new spaces for bicycle parking. And in Peterborough, we have seen multiple, recent additions to the cycling network.

The road to success, said comedian Lily Tomlin, is always under construction. Room for the bicycle is part of that happy path. Build it and they will ride.

Meet FLORC

Some issues live eternal lives. Like unrequited lovers, they manage to reappear, even decades later. Think of Quebec separatism, proposed oil drilling in the Arctic National Wildlife Refuge, a pipeline in the Mackenzie Valley. In my city of Peterborough, too, we have our own timeless issue. We know it as the Parkway.

It seems obvious. Build additional capacity for cars—in my town, a road extension coupled it to the existing thoroughfare, via a bridge through kilometres of greenspace. The benefits could be manifold. Traffic congestion will be eased, travel times shortened, vehicles diverted from residential streets, accidents reduced. Every driver knows the aggravation of road congestion. Even street hockey might benefit with fewer breaks in the action. *Car!*

Indeed, we equate such projects with growth and progress. It has been our formula for prosperity for decades.

But opting for expensive, irreversible change, we need answers to key questions. This is the role of research—to provide evidence and understanding so we can assess the likely consequences of our decisions. Experience from elsewhere can help steer us.

And for the worst commute in the country, I need to look just down the road—to metro Toronto, where the typical round-trip to work is 79 minutes. Every year, that amounts to a staggering 41 work-days per commuter sitting in traffic, $3.3-billion in lost productivity.[1] Drivers tell us commuting to work is less desirable than house-cleaning or the work itself.[2]

Can you build your way out of traffic congestion? As guide for this part of the tour, meet FLORC— affectionately known as the Fundamental Law of Road Congestion.[3] To you, FLORC may be a new acquaintance, but she is well known to transportation researchers. Study after study confirms her stubborn character—that the benefits of road construction are fleeting. Any expected time savings from highway developments soon evaporate as more cars, more trucks, and more trips gobble up the additional capacity.[3-5] The Greater Toronto Area provides a fine example of her handiwork.

Yet we ignore her. We continue to treat traffic congestion as a physical phenomenon—how many vehicles, how far, how fast. We now realize this has been the wrong track. Traffic is not a thing, but a behaviour.[6] Pavement does not merely create space for more efficient movement; it changes driving habits. The lesson from FLORC can be summed up in an analogy: Building more roads to alleviate congestion is like trying to cure obesity by loosening your belt. The cause and remedy are behavioural, more so than physical.

What's more, the effect on nearby streets is marginal, even detrimental. Traffic on neighbourhood routes tends to be the same as before, sometimes higher.

What about quality of life? This is the flip side of new roads. Extensions are often not only roadways, but the elimination of greenspace. And, as research shows, greenspace is vital to human health.

Consider childhood obesity. In just three decades, obesity has doubled among children aged 2–5 and adolescents 12–19, and tripled among children 6–11. It brings risk of diabetes, asthma, hypertension, and emotional distress. Obese children are likely to become obese adults. These are the distant early warnings of a colossal strain on healthcare.

Greenspace is a common sense cure. In an Indianapolis study, for example, neighbourhood greenness was associated with less obesity among children and adolescents. For a typical 16 year-old, living amid parks and outdoor amenities, it means being slimmer–a remarkable 11 pounds less for girls, 13 pounds for boys.[7] Parks, fields, and open spaces invite more physical activity, walking, and play.

And it's not just the kids. The elderly living near greenspace are more active outdoors and report better physical and mental health. Likewise, residents of Tokyo with access to nearby nature showed increased longevity, more activity, and better health.[8] Who is opposed to that?

Finally, we must add up the external costs of expanding roads: A list of ills like higher demand for parking, risk of accidents, noise and air pollution, sprawl, and loss of wildlife habitat. These burdens are shared by all. Rarely do they appear on any balance sheet.

We share a common desire for vibrant, livable cities. The 21st century, however, demands a shift in focus–from mobility to proximity, from `how fast` to `how close`.[6] The road to prosperity is less likely to be found in fresh pavement than in a fresh mindset.

CARIBOU CENTURY

In the span of a typical week, the economic indicators are often headed in opposite directions. Canadians may be increasingly confident, but stock markets take a nosedive. Some industrial countries may pull out of recession, while U. S. retail reports are disappointing. The complexities of the modern economy are clear.

Beyond the headlines, our lives are unfolding in an even more intricate theatre. In nature, it is said, everything is connected to everything else. And it's a formidable task to see though this tangle, to reveal the hidden connections that will govern our future. Nature discloses her secrets with great reluctance.

Some actors are invisible. Carbon dioxide is colourless and odourless, just 0.03 per cent of the atmosphere, but a greenhouse gas. Atmospheric scientists say if its concentrations continue to increase, carbon dioxide has the potential to dramatically alter the 21st-century environment.[1] But other, more hopeful players live in our backyards; we encounter their images every day.

Consider caribou, the northern icon etched into the Canadian imagination. More than a symbol on the 25¢ piece, caribou sustained the emergence of humans 35,000 years ago.[2] People relied on this animal for survival well into the 20th century.

Now, once again, our fortunes and those of caribou are merging.

The woodland caribou, the shy and reclusive

creature of the boreal forest, has a special link to the future. Canada's boreal forest includes some of the most intact forests on the planet—woodlands and peatlands that stretch from Labrador to Yukon. This forest holds more carbon than temperate and tropical forests, including the Amazon, combined—about 27 years' worth of global carbon emissions from burning fossil fuels.[3]

Woodland caribou are seen as an indicator of the forest's healthy functioning. There is remarkable overlap between carbon and caribou. The most carbon-rich stores in the boreal forest are found in the habitat most important to this animal.[3]

But while science has learned much about caribou, we have overlooked an important matter—what we might learn *from* caribou. Perhaps, at this crucial juncture, this animal can point the way to prosperity.

Scale is the new tie that binds us. Here is an animal that needs immense tracts of forest, at least a half-century old.[4] These habitat needs are a challenge, but they are also a call for us to think on a scale that rivals the vast Canadian landscape.

From this vantage point, a new landscape is revealed. We see that half of the woodland caribou range has vanished, swept away by our expanding population and widening ecological footprint.[5-7] But somewhere along the way, our collective memory has also been lost. Although caribou in these forests were as recent as our great-grandparents' day, they now seem as remote to us as the last Ice Age.

As we continue to scrutinize the boreal forest for its resource potential, pressures on woodland caribou will increase. Striking the right balance will test science, our skills at compromise, and our way of thinking.

Familiar arguments may resurface, that conservation will hurt the economy. But our new perspective exposes the old, economy-versus-ecology debate as a false construct and points to the true alternatives: short-term benefits and long-term prosperity. Sustainability means a focus on enduring wealth, especially for future citizens. Rather than an impediment, woodland caribou could be a reassuring sign, a signal of security for us in the long haul.

And here, the news is encouraging. Businesses, from booksellers to paper companies, are adopting policies and practices to help conserve the boreal forest. Governments are responding to the need to protect habitat. The Alberta government, for instance, gave the Lower Athabasca Regional Advisory Council a mandate to identify at least 20 per cent of the region for conservation. Now, as in the past, caribou have a way of bringing people together.

Caribou—lost or conserved? The answer is not strictly a matter of biology. The fate of this species is a record of our values, carved into the landscape and laid bare for succeeding generations. Caribou represent a deep connection to our past and, if we choose, a guidepost for the future. We simply need to adopt a caribou's eye view.

BRING ON THE WHITE, NOT THE BLUES

It is an intrusion. Each year, although we brace for its coming, the arrival is a shock, upsetting routines, travel plans, even our mood. What could cause such trouble? Those incessant Christmas commercials? Shortening day length? Another NHL lock-out? No, this demon is more annoying than the barrage of holiday ads. It is The First Snow.

A new snowfall is an annoyance–turning superhighway into chaos and derailing our just-in-time plans. Canadians may regard themselves as a hardy, northern people, but when it comes to snow, we plow it, shovel it, salt it, and otherwise try to wish it away. Snow is a four-letter word.

But in many regions of the country, the First Snow brings the sparkle of celebration. For northern peoples, it unlocks the world to travel–to hunt, fish, trap, visit neighbouring communities, and be on the land. Winter roads reopen. Old trails are restored; new ones are blazed. In the North, snow is the superhighway.

And to many northern animals, snow is crucial.[1,2] It is the great insulator. And no wonder: 10 cm of snow typically equals 1 cm of water, so new snow is roughly 90% air. At minus 20 degrees Celsius, my students are astonished to see, under a snowy blanket, the ground near freezing. This is where small mammals make a living–a warm, dark, secretive world where they can avoid predators and the extremes of winter.

Even today, our survival may depend on snow. A few years ago, an Ancaster woman, lost in a blizzard, survived 3 days in a farm field wearing little more than a winter jacket–a minor miracle attributed to the insulating effect of snow. The igloo and quinzhee rely on the same principle. These snow houses allowed people to master the northern environment; they still provide the best emergency shelter.

Snow has shaped plants and animals living above its surface, too–from the architecture of tree branches that can withstand the snow load, to the body form of animals traversing a snow-bound forest. Some animals are floaters, like caribou with their large hooves; others are stilters, like moose with their long legs. These creatures are chionophiles–literally, "lovers of snow."

But the snow world is changing. Environment Canada reports springtime Arctic snow cover is vanishing at an alarming rate.[3] Because snow reflects solar energy back into space, better than bare ground, the loss of snow will likely accelerate climate warming. The effects could be global, given Arctic snow and ice represent the world's air-conditioner. Further south, we, too, have a diminishing snowpack–more and more frequently, a Winter that Wasn't. And the shrinking springtime pulse of freshwater is blamed in part for declines in the Great Lakes.[4] In Norway, the legendary cycles of lemmings have ended due to mid-winter thaws that disrupt the snow layer, so pivotal to survival.[5] Because small mammals form the base of the food chain, the effects are likely to ripple throughout the ecosystem.

But losing snow means more than disruption to our economy and environment. Its effects will be cultural. The roots of Canadian culture can be found in wilderness, wildlife, and vast snowy spaces. For the British, the battle of Waterloo may have been won on the playing fields of Eton, but for Canadians, our battles—military, sporting, and otherwise—have been won on a few acres of snow. Our dogged national character was born in a wintery environment. Each year, snow renews our link to the past.

Snow can crystallize our thinking, too. And once again this year, I'll be welcoming those first few delicate flakes—and stand in wonder at the magnificence of countless more to come. Let it snow.

Acknowledgements

Most of these essays represent op-ed pieces that appeared in the *Toronto Star*, *Edmonton Journal*, and (in particular) the *Peterborough Examiner*. I am grateful to the *Examiner*'s managing editors, Jim Hendry and Kennedy Gordon, for the encouragement and the column space. Trent University afforded me the time and the productive physical space to write.

Hidden behind these pages is the Leopold Leadership Program. As a Leopold Fellow, I received training and strategies from the experts on communicating science. It was a career-changer, and I am indebted to the Program's trainers, staff, fellow Fellows, and founders.

Thanks to Vitória Fernandes Felix who assembled the data for Figure 1.

Finally, I give my gratitude and love to Dale—my wife, frontline editor, eternal supporter, and confidante.

ABOUT THE AUTHOR

James A. Schaefer is Professor of Biology at Trent University, a Fellow with the Leopold Leadership Program, and a member of the International Boreal Conservation Science Panel. He has written dozens of technical and popular articles on matters of conservation, ecology, and science. He lives in Peterborough, Ontario.

Notes and References

Introduction

[1] Medawar, P. B., 1984, *The Limits of Science*. Harper & Row, New York.

[2] Lactation, the production of milk in mammary glands, is a defining feature of mammals. The Venetian master, Tintoretto, painted an event in Greek mythology where Zeus descended from Olympus to pluck the infant Hercules from the breast of the goddess Juno (http://www.nationalgallery.org.uk/paintings/jacopo-tintoretto-the-origin-of-the-milky-way). Milk spurts from her breasts and, according to myth, these droplets crystallize to become the start of the galaxy. The painting is physiologically (if not astronomically) correct. It illustrates that milk is under pressure in the breast–called the milk ejection reflex, under control of the hormonal and nervous systems. What's more, the astronomical term "galaxy" is derived from the Greek, *galactos*, meaning "milk". Indeed, our galaxy is known as (what else?) the Milky Way.

[3] The parallels are vividly illustrated on islands, where we can predict both the diversity of species and the diversity of languages from the island's isolation and size. See: Gavin M. C., and N. Sibanda. 2012. The island biogeography of languages. Global Ecology and Biogeography 21: 958-967.

[4] Pimm, S. L., C. N. Jenkins, R. Abell, T. M.

Notes and References

Brooks, J. L. Gittleman, L. N. Joppa, P. H. Raven, C. M. Roberts, and J. O. Sexton. 2014. The biodiversity of species and their rates of extinction, distribution, and protection. Science 344: 1246752. DOI: 10.1126/science.1246752

[5] UNESCO. 2014. Endangered languages. www.unesco.org/new/en/culture/themes/endangered-languages/ [and] www.unesco.org/new/en/culture/themes/endangered-languages/biodiversity-and-linguistic-diversity/

Smaller Bites, Better Taste

Peterborough Examiner, 6 June 2012. http://www.thepeterboroughexaminer.com/2012/06/07/smaller-bites-bring-out-better-taste

[1] For exact figures on obesity, consult, for instance: Bell, J. F., J. S. Wilson, and G. C. Liu. 2008. Neighborhood greenness and 2-year changes in body mass index of children and youth. American Journal of Preventive Medicine 35:547-553.

[2] Leiserowitz, A. A., R. W. Kates, and T. M. Parris. 2006. Sustainability values, attitudes, and behaviors: a review of multinational and global trends. Annual Review of Environment and Resources 31:413-444

[3] Margolis, R., and M. Myrskylä. 2011. A global perspective on happiness and fertility. Population and Development Review 37:29-56.

[4] Galbraith, J. K. 2001. The Essential Galbraith. Mariner Books, Boston. The economist John Kenneth Galbraith (p.111) noted that "Those who

are hungry have a special claim on resources. So do the measures which remedy this privation. For the same reason there is a special case against the luxury consumption of the well-to-do."

The New Environmental Unconsciousness
Peterborough Examiner, 21 April 2011.
[1] Taylor, A. F., F. E. Kuo, and W. C. Sullivan. 2001. Coping with add: The surprising connection to green play settings. Environment and Behavior 33:54-77.
[2] Faber Taylor, A., and F. E. Kuo. 2009. Children with attention deficits concentrate better after walk in the park. Journal of Attention Disorders 12:402-409.
[3] Fuller, R. A., K. N. Irvine, P. Devine-Wright, P. H. Warren, and K. J. Gaston. 2007. Psychological benefits of greenspace increase with biodiversity. Biology Letters 3:390-394.
[4] Kuo, F. E., and W. C. Sullivan. 2001. Environment and crime in the inner city: Does vegetation reduce crime? Environment and Behavior 33:343-367.
[5] Berman, M. G., J. Jonides, and S. Kaplan. 2008. The cognitive benefits of interacting with nature. Psychological Science 19:1207-1212.
[6] In Stone, D. 2006. Sustainable development: Convergence of public health and natural environment agendas, nationally and locally. Public Health 120:1110-1113.
[7] Carbonell, A., and J. M. Gowdy. 2007.

Notes and References

Environmental degradation and happiness. Ecological Economics 60:509-516.
[8] White M., Alcock I., Wheeler B., and Depledge M.H. 2013. Would you be happier living in a greener urban area? A fixed-effects analysis of panel data. Psychological Science 24:920–928.

Taxing Nature's Services
Peterborough Examiner, 6 February 2007.
[1] Dobson, A., D. Lodge, J. Alder, G. S. Cumming, J. Keymer, J. McGlade, H. Mooney, J. A. Rusak, O. Sala, V. Wolters, D. Wall, R. Winfree, and M. A. Xenopoulos. 2006. Habitat loss, trophic collapse, and the decline of ecosystem services. Ecology 87:1915-1924.
[2] Vitousek, P. M., Mooney, H. A., Lubchenco, J. and Melilo, J. M. 1997. Human domination of the earth's ecosystems. Science 277:494-499.
[3] Kremen, C., N. M. Williams, R. L. Bugg, J. P. Fay, and R. W. Thorp. 2004. The area requirements of an ecosystem service: crop pollination by native bee communities in California. Ecology Letters 7:1109-1119.

Aesop's Tortoise
Toronto Star, 18 December 2007.
http://www.thestar.com/opinion/2007/12/18/like_aesops_tortoise_a_slower_approach_is_best.html
[1] Parmesan, C. 1999. Poleward shifts in geographical ranges of butterfly species associated with regional warming. Nature 399:579-583.

Notes and References

[2] May, R. M. 2010. Ecological science and tomorrow's world. Philosophical Transactions: Biological Sciences 365:41-47.
[3] Meyer, S. M. 2004. End of the wild. Boston Review.
http://mitpress.mit.edu/catalog/item/default.asp?ttype=2&tid=10941
[4] Woodruff, D. S. 2001. Declines of biomes and biotas and the future of evolution. Proceedings of the National Academy of Sciences 98:5471-5476.
[5] Monbiot, G. 2006. Heat: How to Stop the Planet from Burning. Doubleday Canada, Toronto.
[6] Storkey, A. 2005. Motorway-based national coach system. www.bepj.org.uk/wordpress/wp-content/2007/03/motorway-based-coach-system.pdf

Pipe Dreams and Future Fortunes

Peterborough Examiner, 30 August 2012.
http://www.thepeterboroughexaminer.com/2012/08/30/pipe-dreams-and-lure-future-fortune
[1] CBC News. 2014. Northern Gateway is not alone: 5 more pipelines to watch.
http://www.cbc.ca/news/politics/northern-gateway-is-not-alone-5-more-pipelines-to-watch-1.2678406

Growing to Extinction

Peterborough Examiner, 2 January 2007.
[1] May, R. M. 2010. Ecological science and tomorrow's world. Philosophical Transactions:

Notes and References

Biological Sciences 365:41-47.

[2] For a US example, see: Trauger, D. L., Czech, B., Erickson, P. R., Garrettson, P. R., Kernohan, B. J. and Miller, C. A. 2003. The relationship of economic growth to wildlife conservation. The Wildlife Society. Technical Review 03-1.

[3] Laliberte, A. S., and W. J. Ripple. 2004. Range contractions of North American carnivores and ungulates. BioScience 54:123-138.

[4] Venter, O., N. N. Brodeur, L. Nemiroff, B. Belland, I. J. Dolinsek, and J. W. A. Grant. 2006. Threats to endangered species in Canada. BioScience 56:903-910.

[5] Davidson, E. A. 2000. You Can't Eat GNP: Economics As If Ecology Mattered. Perseus Books.

[6] Mangel, M., and others. 1996. Principles for the conservation of wild living resources. Ecological Applications 6:338-362.

[7] Stern, N. 2006. Stern Review on The Economics of Climate Change. HM Treasury, London. The Stern report has been influential, but also criticized for the low discount rate (1.4%) used in its projections, compared to William Nordhaus (5.5%), for example. This difference is important; it leads to divergent conclusions about the urgency to address climate change. While discounting cannot be divorced from ethics, some recent literature stresses precaution as a response to averting unfavourable surprises, social discount rates that approach zero for the distant future, and the uncertainty about the discount rate yielding a higher net present value.

Notes and References

These findings lean toward Stern's conclusions. Indeed, some economists have acknowledged Stern "may be right ... for the wrong reasons" (in Cole, D. H. 2008, The *Stern Review* and its critics: implications for the theory and practice of benefit-cost analysis, Natural Resources Journal 48:53-90). See also: Arrow, K. et al., 2012, Determining benefits and costs for future generations, Science 341:349-350; Gowdy, J. et al., 2013, The evolution of hyperbolic discounting: Implications for truly social valuation of the future, Journal of Economic Behavior & Organization 90S: S94– S104; Hayward, T. H., 2012, Climate change and ethics, Nature Climate Change, doi: 10.1038/NCLIMATE1615; Quiggin, J., 2007, Complexity, climate change and the precautionary principle, Environmental Health 7:15–21.

[8] Lubchenco, J. 1998. Entering the century of the environment: A new social contract for science. Science 279:491-497.

To Be a Better Ancestor
Peterborough Examiner, 22 January 2015
http://www.thepeterboroughexaminer.com/2015/01/22/to-be-a-better-ancestor--n-our-changing-seasons-a-growing-chorus-of-economists-and-scientists-is-calling-for-a-price-on-carbon

[1] Jedwab, J. 2003. Participating in Sports and Fitness Activities in Canada. www.acs-aec.ca/pdf/polls/Poll18.pdf

[2] Dimitri, N., and J. van Eijck. 2010. Time

Notes and References

discounting and time consistency. In: J. van Eijck and R. Verbrugge (Eds.) Games, Actions, and Social Software. homepages.cwi.nl
/~jvc/papers/11/pdfs/tdChapter.pdf

[3] Goulder, L. H. and R. N. Stavins. 2002. An eye to the future. Nature 419: 673-674.

[4] Clapham, P. J., A. Aguilar, and L. Hatch. 2008. Determining scales for management units. Marine Mammal Science 24: 183–201.

[5] Caughley, G., and A. Gunn. 1996. Conservation Biology in Theory and Practice. Blackwell Science, Cambridge, MA.

[6] Hayward, T. 2012. Climate change and ethics. Nature Climate Change DOI: 10.1038/NCLIMATE1615.

[7] Gowdy, J., J. B. Rossser, and J. Roy. 2013. The evolution of hyperbolic discounting: Implications for truly social valuation of the future. Journal of Economic Behavior & Organization 90S: S94– S104

[8) A great question posed (with respect to Great Bear Rainforest) by Fisher, B, R. Naidoo and T. Ricketts. 2015. A Field Guide to Economics for Conservationists. Roberts & Company Publishers Greenwood Village, CO.

[9] Simpson, J., M. Jaccard, and N. Rivers. 2007. Hot Air–Meeting Canada's Climate Change Challenge. McClelland & Stewart, Toronto.

[10] Weaver, A. 2008. Keeping Our Cool: Canada in a Warming World. Viking Canada, Toronto.

Notes and References

Gambling Without Spin
Peterborough Examiner, 5 May 2013.
http://www.thepeteroroughexaminer.com/2013/0
5/23/fortunately-blind-chance-wont-decide-if-
casino-arrives

[1] Rosenthal, J. S. 2006. Struck By Lightning: The Curious World of Probabilities. HarperCollins Canada, Toronto.

[2] Croson, R., and J. Sundali. 2005. The gambler's fallacy and the hot hand: empirical data from casinos. Journal of Risk and Uncertainty 30:195-209.

[3] Harrigan, K. A. and M. Dixon. 2009. PAR Sheets, probabilities, and slot machine play: Implications for problem and non-problem gambling. Journal of Gambling Issues 23:81-110.

[4] Wiebe, J., Mun, P. & Kauffman, N. 2006. Gambling and Problem Gambling in Ontario 2005. Toronto: Responsible Gambling Council.

Are You Scientifically Literate?
Chance News 94 [online]
http://test.causeweb.org/wiki/chance/index.php/C
hance_News_94

[1] Toronto Star, 11 August 2009. Scientific literacy quiz.
http://www.thestar.com/life/2009/08/11/scientific
_literacy_quiz.html

[2] Runté, R. 1995. Basic rules for taking a multiple choice test.
http://www.uleth.ca/edu/runte/tests/take/mc/ho
w.html

Notes and References

[3] For the statistically inclined, you can apply a binomial test, which reveals the probability (P) of obtaining such a result (or better) due to sheer luck. The result: $P = 0.040$ (1-tailed). It's very unlikely (1 chance in 25) that random picks would result in such a good score.

[4] Binomial test, $P = 0.00039$ (1-tailed). It's extremely unlikely (roughly 1 chance in 2500) that random picks would give such a good score.

Variation, Meet Variety

Peterborough Examiner, 13 October 2011.

[1] Garber, N. J. and R. Gadiraju. 1989. Factors affecting speed variance and its influence on accidents. Transportation Research Record 1213:64.

[2] Mussa, R., and V. Muchuruza. 2006. Safety analysis of Florida rural interstate freeway travel in relation to the 65 km/h (40 mi/hr) minimum speed regulation. Journal of Transportation Engineering 132:699-707.

[3] Hamel, K. A., N. Okita, J. S. Higginson, and P. R. Cavanagh. 2005. Foot clearance during stair descent: effects of age and illumination. Gait & Posture 21:135-140.

[4] Dall, S. R. X., and I. L. Boyd. 2004. Evolution of mammals: Lactation helps mothers to cope with unreliable food supplies. Proceedings: Biological Sciences 271:2049-2057.

[5] Bates, B.C., Z. W. Kundzewicz, S. Wu and J. P. Palutikof, Eds. 2008. Climate change and water. IPCC Technical Paper VI. IPCC Secretariat,

Notes and References

Geneva.

[6] Salinger, M. J. 2005. Climate variability and change: past, present and future–an overview. Climatic Change 70:9-29.

[7] Morin, E. 2008. Evidence for declines in human population densities during the early Upper Paleolithic in western Europe. Proceedings of the National Academy of Sciences 105:48-53.

[8] Piñeiro, G., E. G. Jobbágy, J. Baker, B. C. Murray, and R. B. Jackson. 2009. Set-asides can be better climate investment than corn ethanol. Ecological Applications 19:277-282.

[9] Steiner, C. F., Z. T. Long, J. A. Krumins, and P. J. Morin. 2006. Population and community resilience in multitrophic communities. Ecology 87:996-1007.

Your Days Are Numbered

Peterborough Examiner, 18 July 2013.
http://www.thepeteroroughexaminer.com/2013/07/18/cant-figure-the-probabilities-your-days-are-numbered

[1] Duhigg, C. 2009. What does your credit-card company know about you? New York Times, 12 May 2009.
http://www.nytimes.com/2009/05/17/magazine/17credit-t.html?pagewanted=all&_r=0

[2] Monbiot, G. 2006. Heat: How to Stop the Planet from Burning. Doubleday Canada, Toronto.

[3] Mattern, K. D., Burrus, J., & Shaw, E. 2009. When both the skilled and unskilled are unaware: Consequences for academic performance. Self and

Notes and References

Identity 9: 129-141.
[4] Sunstein, C. R., & R. H. Thaler. 2009. Nudge: Improving Decisions about Health, Wealth, and Happiness. Penguin Group, Toronto.
[5] Rosenthal, J. S. 2006. Struck By Lightning: The Curious World of Probabilities. HarperCollins Canada, Toronto.

Uncertainty? Bet on it
Peterborough Examiner, 21 December 2009.
[1] Giles, J. 2005. Climate sceptics place bets on world cooling down. Nature 436:897.
[2] Rosenthal, J. S. 2006. Struck By Lightning: The Curious World of Probabilities. HarperCollins Canada, Toronto.
[3] Keller, G., T. Adatte, W. Stinnesbeck, M. Rebolledo-Vieyra, J. U. Fucugauchi, U. Kramar, D. Stüben, and W. J. Morgan. 2004. Chicxulub impact predates the K-T Boundary mass extinction. Proceedings of the National Academy of Sciences 101:3753-3758.
[4] IPCC, 2013. Climate Change 2013: The Physical Science Basis. Contribution of Working Group I to the Fifth Assessment Report of the Intergovernmental Panel on Climate Change [Stocker, T.F., D. Qin, G.-K. Plattner, M. Tignor, S.K. Allen, J. Boschung, A. Nauels, Y. Xia, V. Bex and P.M. Midgley (eds.)]. Cambridge University Press, Cambridge, United Kingdom and New York, NY, USA.
[5] Schneider S. 2009. The worst-case scenario.

Nature 458: 1104-1105.
[6] Smith, J. B., S. H. Schneider, M. Oppenheimer, G. W. Yohe, W. Hare, M. D. Mastrandrea, P. Anand, I. Burton, J. Corfee-Morlot, C. H. D. Magadza, H. M. Füssel, A. B. Pittock, A. Rahman, A. Suarez, and J. P. v. Ypersele. 2009. Assessing dangerous climate change through an update of the Intergovernmental Panel on Climate Change (IPCC) "Reasons for Concern". Proceedings of the National Academy of Sciences 106:4133-4137.

The Tale of Lord and Lady, Minus a Few Characters
Peterborough Examiner, 2 June 2011.
[1] Ehrlich, P. R., and A. H. Ehrlich. 2004. One with Nineveh: Politics, Consumption, and the Human Future. Island Press, Washington.
[2] Homer-Dixon, H. 2006. The Upside of Down: Catastrophe, Creativity, and the Renewal of Civilization. Alfred A. Knopf Canada, Toronto.
[3] Wallace, A. R. 1863. On the physical geography of the Malay Archipelago. Journal of the Royal Geographical Society of London 33:217-234.
[4] Rockström, J., and others. 2009. A safe operating space for humanity. Nature 461:472-475.

Moon Shot or Pie in the Sky?
Peterborough Examiner, 15 March 2012.
[1] Cohen, J. E., and D. Tilman. 1996. Biosphere 2 and biodiversity–the lessons so far. Science 274:1150-1151.

Notes and References

[2] Walford, R. L. 2002. Biosphere 2 as voyage of discovery: the serendipity from inside. BioScience 52:259-263.

Life is a Highway
[1] Baldwin, A., T. Harris, and G. Davies. 2008. Look right! A retrospective study of pedestrian accidents involving overseas visitors to London. Emergency Medicine Journal 25:843-846.
[2] IFrom World Standards (http://www.worldstandards.eu/cars/list-of-left-driving-countries/), I computed that 58% (59/101) of island nations and territories drove on the left, compared to just 15% (24/161) of those on the mainland.
[3] The Guardian. 2009. Samoa switches smoothly to driving on the left. http://www.theguardian.com/world/2009/sep/08/samoa-drivers-switch-left
[4] Rogers, L. J. 2009. Hand and paw preferences in relation to the lateralized brain. Philosophical Transactions: Biological Sciences 364:943-954.
[5] Grant, R. J., V. F Colenbrander, & J. L. Albright. 1990. Effect of particle size of forage and rumen cannulation upon chewing activity and laterality in dairy cows. Journal of Dairy Science 73: 3158-3164.
[6] Frasnelli E. 2013. Brain and behavioral lateralization in invertebrates. Frontiers in Psychology 4: 939.
[7] Ströckens, F., O. Güntürkün, and S.

Notes and References

Ocklenburg. 2012. Limb preferences in non-human vertebrates. Laterality: Asymmetries of Body, Brain and Cognition 18:536-575.

Expect Delays, in Slow Motion
Toronto Star, 4 October 2007.
http://www.thestar.com/opinion/columnists/2007/10/04/climate_dominoes_tumble_slowly.html
[1] IPCC, 2013. Climate Change 2013: The Physical Science Basis. Contribution of Working Group I to the Fifth Assessment Report of the Intergovernmental Panel on Climate Change [Stocker, T.F., D. Qin, G.-K. Plattner, M. Tignor, S.K. Allen, J. Boschung, A. Nauels, Y. Xia, V. Bex and P.M. Midgley (eds.)]. Cambridge University Press, Cambridge, United Kingdom and New York, NY, USA.
[2] Friedlingstein, P. and S. Solomon. 2005. Contributions of past and present human generations to committed warming caused by carbon dioxide. Proceedings of the National Academy of Sciences 102:10832-10836
[3] Vors, L. S., J. A. Schaefer, B. A. Pond, A. R. Rodgers, and B. R. Patterson. 2007. Woodland caribou extirpation and anthropogenic landscape disturbance in Ontario. Journal of Wildlife Management 71:1249-1256
[4] Kolbert, E. 2007. Field Notes from a Catastrophe: Man, Nature, and Climate Change. Bloomsbury Publishing, New York.
[5] Dombeck, M. 2001. The BIG TEN Public Land

Notes and References

Conservation Challenges For a New Century: Where do we go from here?
http://forestry.berkeley.edu/lectures/albright/2001dombeck.html
[6] Davis, S. 1987. Future Perfect. Addison Wesley Publishing.

A Brave New World for Birds and Beasts
Peterborough Examiner, 25 August 2011.
[1] Darwin, C. 2001. The Voyage of the Beagle: Journal of Researches into the Natural History and Geology of the Countries Visited during the Voyage of the H.M.S. *Beagle* Round the World. Modern Library, New York.
[2] McKinney, M. L. 2006. Urbanization as a major cause of biotic homogenization. Biological Conservation 127:247-260.
[3] Hulme, P. E. 2009. Trade, transport and trouble: managing invasive species pathways in an era of globalization. Journal of Applied Ecology 46:10-18.
[4] Pimentel, D., R. Zuniga, and D. Morrison. 2005. Update on the environmental and economic costs associated with alien-invasive species in the United States. Ecological Economics 52:273-288.
[5] Simberloff, D., J. L. Martin, P. Genovesi, V. Maris, D. A. Wardle, J. Aronson, F. Courchamp, B. Galil, E. García-Berthou, M. Pascal, P. Pyšek, R. Sousa, E. Tabacchi, and M. Vilà. 2013. Impacts of biological invasions: what's what and the way forward. Trends in Ecology & Evolution 28:58-66.

Notes and References

[6] Vitousek, P. M., C. M. D'Antonio, L. L. Loope, M. Rejmanek, and R. Westbrooks. 1997. Introduced species: a significant component of human-caused global change. New Zealand Journal of Ecology 21:1-16.

Good Advice Is a Science
Peterborough Examiner, 24 January 2013.
[1] Hall, S. S. 2011. Scientists on trial: At fault? Nature 477:264-269.
[2] Minns, C. K. 1995. Allometry and home range size in lake and river fishes. Canadian Journal of Fisheries and Aquatic Sciences 52: 1499-1508.
[3] Arditti, J., J. Elliott, I. J. Kitching, and L. T. Wasserthal. 2012. 'Good Heavens what insect can suck it'–Charles Darwin, *Angraecum sesquipedale* and *Xanthopan morganii praedicta*. Botanical Journal of the Linnean Society 169:403-432.
[4] Based on points scored during the 2004-05 season by 55 players from NHL teams in Montreal, Ottawa, Toronto, Calgary and Edmonton, I calculated that the Hockey Scouting Report provided a better prediction than past performance in only 36% (20/55) of cases.
[5] Schneider, S. 2011. Understanding and solving the climate change problem.
http://stephenschneider.stanford.edu/Climate/Climate_Impacts/Impacts.html
[6] Friedman, S. M., S. Dunwoody, and C. L. Rogers, editors. 1999. Communicating Uncertainty: Media Coverage of New and Controversial Science.

Notes and References

Lawrence Erlbaum Associates, London.

The Measure of Things
Peterborough Examiner, 19 June 2008.
[1] Allen, T. F. H. and T. W. Hoekstra, 1992. Toward a Unified Ecology. Columbia University Press, New York.
[2] Rettie, W. J., and F. Messier. 2000. Hierarchical habitat selection by woodland caribou: its relationship to limiting factors. Ecography 23:466-478.
[3] Simberloff, D., J. L. Martin, P. Genovesi, V. Maris, D. A. Wardle, J. Aronson, F. Courchamp, B. Galil, E. García-Berthou, M. Pascal, P. Pyšek, R. Sousa, E. Tabacchi, and M. Vilà. 2013. Impacts of biological invasions: what's what and the way forward. Trends in Ecology & Evolution 28:58-66.
[4] Cumming, G. S., D. H. M. Cumming, and C. L. Redman. 2006. Scale mismatches in social-ecological systems: causes, consequences, and solutions . Ecology and Society 11(1): 14.
[5] Rodríguez, J. P., T. D. Beard, Jr., E. M. Bennett, G. S. Cumming, S. Cork, J. Agard, A. P. Dobson, and G. D. Peterson. 2006. Trade-offs across space, time, and ecosystem services. Ecology and Society 11(1): 28.
http://www.ecologyandsociety.org/vol11/iss1/art28/

Evidence Means a Sporting Chance for All
Peterborough Examiner, 19 June 2014.

Notes and References

http://www.thepeteroroughexaminer.com/2014/06/19/a-sporting-chance-for-all-of-us
[1] House of Commons Standing Committee on Canadian Heritage. 1998. Sport in Canada: Everybody's Business. http://www.parl.gc.ca/HousePublications/Publication.aspx?DocId=1031530&Language=E&Mode=1&Parl=36&Ses=1&File=18#partI
[2] Frank, R. A., J. W. Roy, G. Bickerton, S. J. Rowland, J. V. Headley, A. G. Scarlett, C. E. West, K. M. Peru, J. L. Parrott, F. M. Conly, and L. M. Hewitt. 2014. Profiling oil sands mixtures from industrial developments and natural groundwaters for source identification. Environmental Science & Technology 48:2660-2670.
[3] International Joint Commission. 2013. Assessment of Progress Made Towards Restoring and Maintaining Great Lakes Water Quality Since 1987. Sixteenth Biennial Report on Great Lakes Water Quality. ISSN: 0736-8410.

Of Microbes and Men
Peterborough Examiner, 27 February 2014.
http://www.thepeteroroughexaminer.com/2014/02/27/of-microbes-and-men
[1] Spellberg, B., R. Guidos, D. Gilbert, J. Bradley, H. W. Boucher, W. M. Scheld, J. G. Bartlett, J. Edwards, and the Infectious Diseases Society of America. 2008. The epidemic of antibiotic-resistant infections: a call to action for the medical community from the Infectious Diseases Society of America.

Notes and References

Clinical Infectious Diseases 46:155-164.
[2] Understanding Evolution. 2014. University of California Museum of Paleontology. 22 August 2008 http://evolution.berkeley.edu/
[3] Coltman, D. W., P. O'Donoghue, J. T. Jorgenson, J. T. Hogg, C. Strobeck, and M. Festa-Bianchet. 2003. Undesirable evolutionary consequences of trophy hunting. Nature 426:655-658.
[4] Gee, H., R. Howlett, and P. Campbell. 2009. 15 evolutionary gems. Nature doi:10.1038/nature07740.
[5] Kuparinen, A., and J. Merila. 2007. Detecting and managing fisheries-induced evolution. Trends in Ecology and Evolution 22:652-659.

Translation Is a Tricky Business

Peterborough Examiner, 5 December 2013.
http://www.thepeterboroughexaminer.com/2013/12/05/translating-science-into-english
[1] Wade, N. 2011. Phonetic clues hint language is africa-born. New York Times, 14 April 2011. http://www.nytimes.com/2011/04/15/science/15language.html?_r=0
[2] Baron, N. 2010. Escape from the Ivory Tower: A Guide to Making Your Science Matter. Island Press, Washington.
[3] Dean, C. 2009. Am I Making Myself Clear? A Scientist's Guide to Talking to the Public. Harvard University Press, Cambridge, Massacheusetts.
[4] Heath, C. and D. Heath. 2007. Made to Stick:

Notes and References

Why Some Ideas Survive and Others Die. Random House, New York.

[5] Novacek, M. J. 2008. Engaging the public in biodiversity issues. Proceedings of the National Academy of Sciences 105:11571-11578.

[6] Schaefer, J. A., and P. Beier. 2013. Going public: Scientific advocacy and North American wildlife conservation. International Journal of Environmental Studies 70:429-437.

Looking Past the Coin Flip

Peterborough Examiner, 5 September 2013. http://www.thepeterboroughexaminer.com/2013/09/05/science-means-looking-past-the-coin-flip

[1] This lesson in skepticism I adapted from Dr. Josh Tenebaum, cognitive scientist at the Massachusetts Institute of Technology, as described in Cory Dean's (2009) book, *Am I Making Myself Clear? A Scientist's Guide to Talking to the Public.*

[2] Sagan, C. 1995. The Demon Haunted World: Science as a Candle in the Dark. Ballantine Books, New York.

[3] Baum, J. K., R. A. Myers, D. G. Kehler, B. Worm, S. J. Harley, and P. A. Doherty. 2003. Collapse and conservation of shark populations in the northwest Atlantic. Science 299:389-392.

[4] Biggs, D., F. Courchamp, R. Martin, and H. P. Possingham. 2013. Legal trade of Africa's rhino horns. Science 339:1038-1039.

[5] Laburn, H. P. and D. Mitchell. 1997. Extracts of rhinoceros horn are not antipyretic in rabbits.

Notes and References

Journal of Basic and Clinical Physiology 8:1-11.
[6] Cassileth, B. R., and I. R. Yarett. 2012. Cancer quackery: The persistent popularity of useless, irrational 'alternative' treatments. Oncology 26:754-758.
[7] Cook, J. and others. 2013. Quantifying the consensus on anthropogenic global warming in the scientific literature. Environmental Research Letters 8:024024.

A Little Feedback on your Weight
Peterborough Examiner, 6 November 2014.
http://www.thepeterboroughexaminer.com/2014/11/06/a-little-feedback-on-your-weight
[1] Cuevas, A. M., M. M. Farias, and F. Rodriguez. 2011. Set-point theory and obesity. Metabolic Syndrome and Related Disorders 9:85-89.
[2] Hill, J. O., H. R. Wyatt, and J. C. Peters. 2012. Energy balance and obesity. Circulation 126:126-132.
[3] Dasgupta, P. 2001. Human well-being and the natural environment. Oxford University Press.
[4] Haushofer, J. and E. Fehr. 2014. On the psychology of poverty. Science 344: 862-867.
[5] Brumfiel, G. 2009. Kaputnik chaos could kill Hubble. Nature doi:10.1038/457940a
[6] Schiermeier, D. 2013. Wild weather can send greenhouse gases spiralling. Nature 496:137.

Build it and They Will Bike
Peterborough Examiner, 21 March 2013.

Notes and References

http://www.thepeteroroughexaminer.com/2013/0 3/21/build-it-and-they-will-bike

[1] Natural Resources Canada. 2010. 2008 Canadian vehicle survey–update report. http://oee.nrcan.gc.ca/publications/statistics/cvs08 /pdf/cvs08.pdf

[2] U.S. Department of Transportation. 2012. November 2012 traffic volume and trends. http://www.fhwa.dot.gov/policyinformation/travel _monitoring/12novtvt/index.cfm

Ipsos Reid. 2009. City of Toronto Cycling Study Tracking Report (1999 and 2009). City-of-Toronto-Cycling-Study-Tracking-Report-1999-and-2009.pdf

Meet FLORC

Peterborough Examiner, 23 April 2011 and 12 November 2013.

http://www.thepeteroroughexaminer.com/2013/1 1/13/parkway-new-roads-can-get-busy-too

[1] Turcotte, M. 2006. The time it takes to get to work and back. Statistics Canada Catalogue no. 89-622-XIE, General Social Survey on Time Use: Cycle 19.

http://publications.gc.ca/Collection/Statcan/89-622-X/89-622-XIE2006001.pdf

[2] Kahneman, D., A. B. Krueger, D. A. Schkade, N. Schwarz, and A. A. Stone. 2004. A survey method for characterizing daily life experience: The day reconstruction method. Science 306:1776-1780.

[3] Duranton, G., and M. A. Turner. 2011. The fundamental law of road congestion: evidence from

Notes and References

US cities. American Economic Review 101:2616-2652.

[4] Litman, T. 2010. Generated traffic and induced travel. Implications for Transport Planning. Victoria Transport Policy Institute.
www.vtpi.org/gentraf.pdf

[5] Næss, P., M. S. Nicolaisen and A. Strand. 2012. Traffic forecasts ignoring induced demand: a shaky fundament for cost-benefit analyses. European Journal of Transport and Infrastructure Research 12:291-309.

[6] Ladd, B. 2012. "You can't build your way out of congestion."–Or can you? The Planning Review 48:16-23.

[7] Bell, J. F., J. S. Wilson, and G. C. Liu. 2008. Neighborhood greenness and 2-year changes in body mass index of children and youth. American Journal of Preventive Medicine 35:547-553.

[8] In Stone, D. 2006. Sustainable development: Convergence of public health and natural environment agendas, nationally and locally. Public Health 120:1110-1113.

Caribou Century

Edmonton Journal, 28 August 2009.
http://www2.canada.com/edmontonjournal/news/opinion/story.html?id=39747632-d7a5-4ea1-9d05-7c7e83654bd1

[1] IPCC, 2013. Climate Change 2013: The Physical Science Basis. Contribution of Working Group I to the Fifth Assessment Report of the

Notes and References

Intergovernmental Panel on Climate Change [Stocker, T.F., D. Qin, G.-K. Plattner, M. Tignor, S.K. Allen, J. Boschung, A. Nauels, Y. Xia, V. Bex and P.M. Midgley (eds.)]. Cambridge University Press, Cambridge, United Kingdom and New York, NY, USA.

[2] Morin, E. 2008. Evidence for declines in human population densities during the early Upper Paleolithic in western Europe. Proceedings of the National Academy of Sciences 105:48-53.

[3] Carlson, M., J. Wells, and D. Roberts. 2009. The carbon the world forgot: Conserving the capacity of Canada's boreal forest region to mitigate and adapt to climate change. Boreal Songbird Initiative and Canadian Boreal Initiative, Seattle, WA, and Ottawa.

[4] Schaefer, J. A., and W. O. Pruitt, Jr. 1991. Fire and woodland caribou in southeastern Manitoba. Wildlife Monographs 116:1-39.

[5] Schaefer, J. A. 2003. Long-term range recession and the persistence of caribou in the taiga. Conservation Biology 17:1435-1439.

[6] Hervieux, D., M. Hebblewhite, N. J. DeCesare, M. Russell, K. Smith, S. Robertson, and S. Boutin. 2013. Widespread declines in woodland caribou (*Rangifer tarandus caribou*) continue in Alberta. Canadian Journal of Zoology 91:872-882.

[7] Ray, J. C. and M. Hummel. 2008. Caribou and the North: a Shared Future. Dundurn Press, Toronto.

Notes and References

Bring on the White, Not the Blues
Peterborough Examiner, 22 November 2012.
http://www.thepeteroboroughexaminer.com/2012/1
1/22/bring-on-the-white-not-the-blues
[1] Pruitt, W. O., Jr. 1960. Animals in the snow. Scientific American 202:60-68.
[2] Pruitt, W. O., Jr. 2005. Why and how to study a snowcover. Canadian Field-Naturalist 119:118-128.
[3] Derksen, C., and R. Brown. 2012. Spring snow cover extent reductions in the 2008-2012 period exceeding climate model projections. Geophysical Research Letters 39:L19504.
[4] Cruce, T., & E. Yurkovich. 2011. Adapting to climate change: A planning guide for state coastal managers—a Great Lakes supplement. NOAA Office of Ocean and Coastal Resource Management, Silver Spring, MD.
http://coastalmanagement.noaa.gov/climate/docs/adaptationgreatlakes.pdf
[5] Kausrud, K. L., A. Mysterud, H. Steen, J. O. Vik, E. Ostbye, B. Cazelles, E. Framstad, A. M. Eikeset, I. Mysterud, T. Solhoy, and N. C. Stenseth. 2008. Linking climate change to lemming cycles. Nature 456:93-97.

CPSIA information can be obtained at www.ICGtesting.com
Printed in the USA
BVOW04s0905150715

408722BV00001B/1/P